TM 5-277

FIXED STEEL PANEL BRIDGE BAILEY TYPE

TECHNICAL MANUAL

RESTRICTED *Dissemination of restricted matter.*—The information contained in restricted documents and the essential characteristics of restricted material may be given to any person known to be in the service of the United States and to persons of undoubted loyalty and discretion who are cooperating in Government work, but will not be communicated to the public or to the press except by authorized military public relations agencies. (See also par. 18b, AR 380–5, 28 Sep 1942.)

BY *WAR DEPARTMENT* · *15 SEPTEMBER 1943*

DISCLAIMER:

This manual is sold for historic research purposes only, as an entertainment. It contains obsolete information and is not intended to be used as part of an actual operation or maintenance training program. No book can substitute for proper training by an authorized instructor.

WAR DEPARTMENT TECHNICAL MANUAL

TM 5-277

FIXED STEEL
PANEL BRIDGE
BAILEY TYPE

WAR DEPARTMENT • 15 SEPTEMBER 1943

UNITED STATES GOVERNMENT PRINTING OFFICE
WASHINGTON 1943

FIXED STEEL PANEL BRIDGE

BAILEY TYPE

Contents

Figure 1.—Medium tank crossing 100-foot double-single bridge.

GENERAL

1. **PURPOSE OF BRIDGE.**—The fixed steel panel bridge, Bailey type, is designed to carry all military loads. It can span distances up to 180 feet (fig. 1) and can be assembled to meet varying conditions of span and load.

2. **GENERAL DESIGN.**—*a. Main supporting members.*— The bridge is supported by two main trusses formed from 10-foot truss sections called panels. A bay consists of two parallel truss sections supporting transverse members, called transoms, and the tringers and decking. The roadway, supported by stringers carried on the transoms, has a clear width of 10 feet 9 inches. On each side of the bridge can be added a footwalk 2 feet 6 inches wide, carried on the transoms outside the main trusses.

   b. Materials and method of fabrication.— The chess and ribands (see par. 19) are wood. All other parts are steel with welded joints.

   c. Types of bridges.— The strength of the main trusses may be increased by adding extra panels alongside, by adding an extra story, or both. For example, each side of each bay of a double-truss, double-story—called double-double—bridge (fig. 2) consists of two paralled panels, with two others above them. Only the five following types (fig. 3) may be used:

Type	Usual nomenclature	Abbreviation
Single-truss, single story	Single-single	SS
Double-truss, single story	Double-single	DS
Triple-truss, single story	Triple-single	TS
Double-truss, double story	Double-double	DD
Triple-truss, double story	Triple-double	TD

1

A single-truss, double-story bridge never is built, having insufficient lateral bracing. The capacity of an existing bridge can be increased by adding panels to each bay.

d. Component parts.— Construction of the bridge requires 29 different kinds of parts and 22 different kinds of tools. These are described in detail in chapter 2. Figures 4 and 5 show the relative positions of the major parts.

3. CAPACITY.—*a. General.*— Table I gives the posted capacities of panel bridges of various types of construction and different spans. It also shows which typical vehicles may cross.

b. How to post the bridge.— Posted capacities—values at which the bridges normally are rated—are shown at the top of table I. For example, a triple-truss, single-story bridge with a span of 140 feet is posted at 18 tons.

c. How to use posted values.— See FM 5-10.

4. SIZE OF WORKING PARTY REQUIRED.—The size of working party required for various types of panel bridge is shown in table II.

Figure 2.—Double-double bridge.

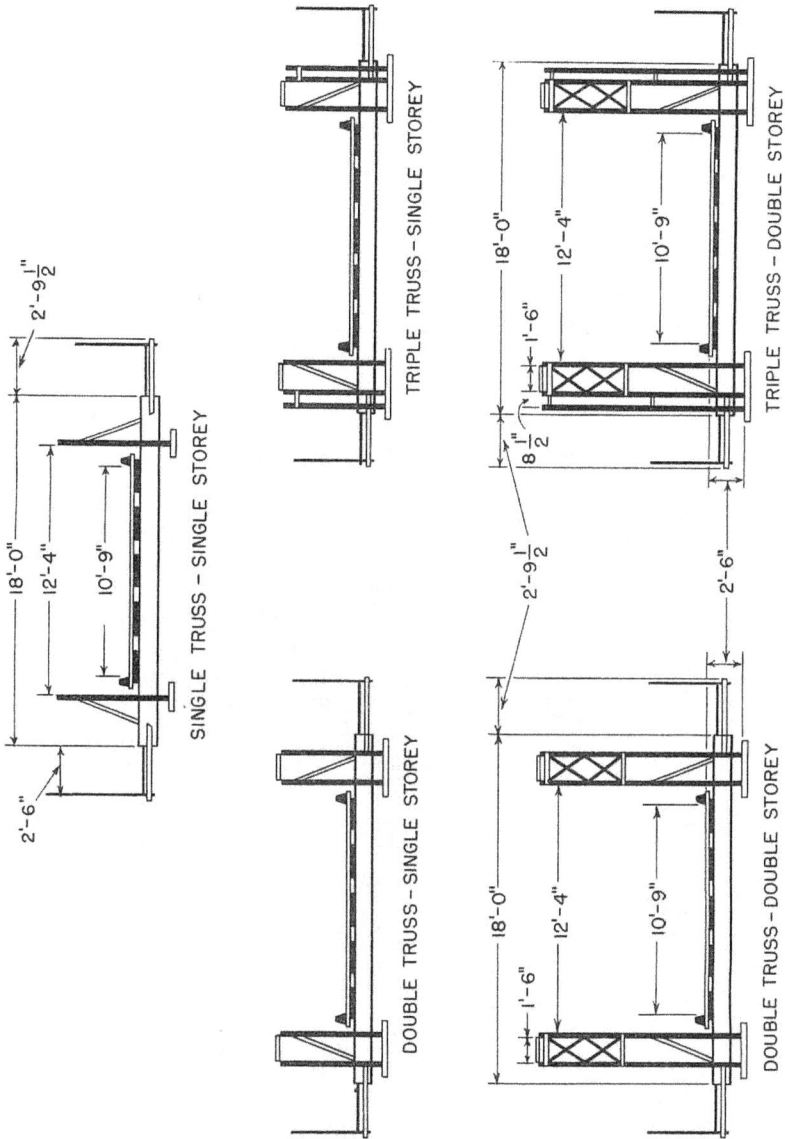

Figure 3.—Types of construction.

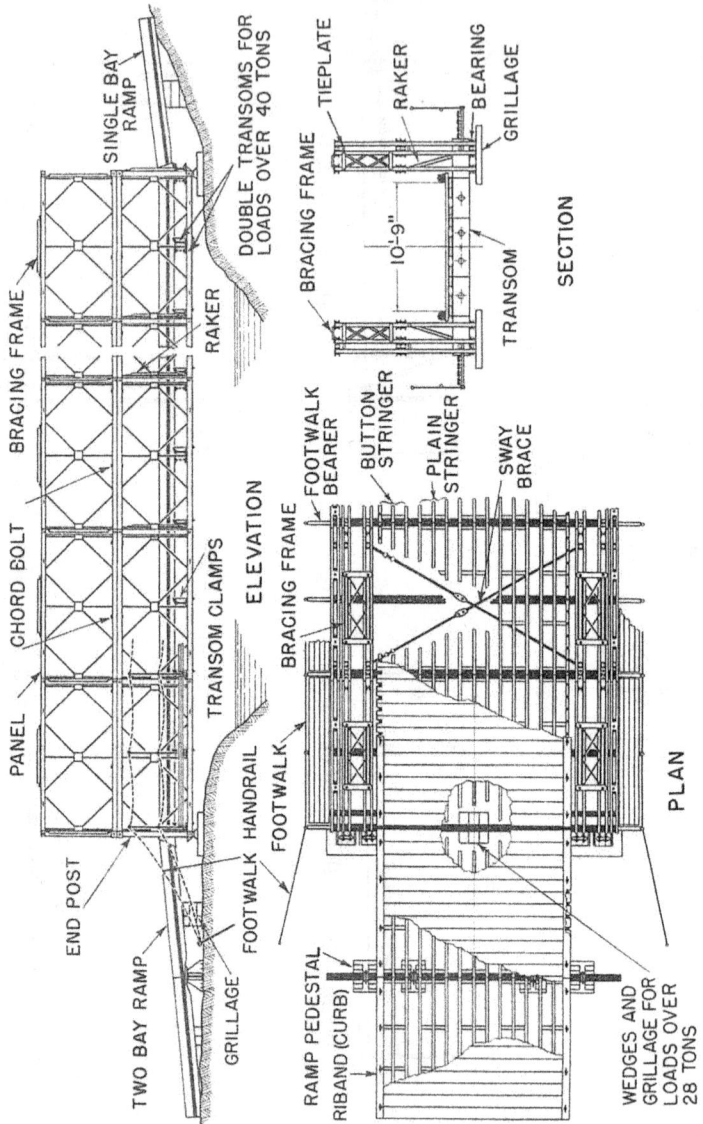

Figure 4.—Triple-double bridge—nomenclature.

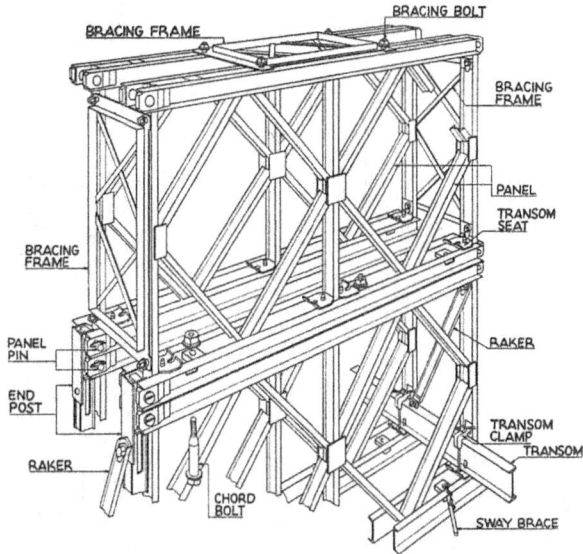

Figure 5.—End bay panels of double-double bridge—nomenclature.

5. ESTIMATED TIME FOR CONSTRUCTION.—*a.* Tables III*a* and III*b* show estimated construction time for the various span lengths of different types of bridges.

b. The estimated times given in tables III*a* and III*b* assume a favorable construction site, trained personnel, equipment stacked at site, and footwalks omitted. Add 30 minutes for unloading from trucks if a separate unloading detail is available; if not available, add 1 to 2½ hours according to type of bridge. Fatigue, poor weather, and enemy activity also will lengthen construction time.

6. COMPOSITION AND ASSIGNMENT OF EQUIPMENT.—*a. Bridge unit.*— A single unit of the equipment, enough for 150 feet of double-double or for one 80-foot and one 70-foot triple-single bridge, consists of the parts shown in table IV.

b. Issue.— See T B/A or FM 5-35.

5

VEHICLE	WT.CLASS-TONS	SS			DS							TS							DD							TD					
		SPAN OF BRIDGE IN FEET																														
		60	50	30	120	110	100	80	60	50	40	140	130	120	110	90	80	70	160	150	140	130	110	90	80	170	160	150	130	120	110
		POSTED CAPACITY IN TONS																														
		28	35	45	16	20	23	33	50	60	70	18	22	25	30	41	48	60	21	25	30	35	45	53	61	19	23	29	35	49	57	65
Truck, 1½-T, w/1-T tlr.	6																														
Truck, 1½-T, w/105-mm How.	6																														
Tractor, D-4, w/dozer	7																														
Car, armored, light, M8	8																														
Truck, 2½-T, w/1-T tlr.	9																														
Truck, 2½-T, w/105-mm How.	9																														
Car, half track, M2	9																														
Other vehicles under 10 tons																															
Grader, road, mtzd. (Engr.)	11																														
Truck, 4-T, wrecker	11																														
Tank, light, M2A4	12																														
Truck, 2½-T, w/155-mm How., carr. M1	11																														
Crane, trk-mtd. (Engr.)	12																														
Truck, 4-T, cargo (same as distributor, water)	13																														
Truck, 4-T, ponton	13																														
Tank, light, M3	14																														
Trk-tractor, 4-5T, w/semi-tlr., fuel serv., F-2 (AC)	12																														
Tractor, D-7, w/dozer	15																														
Truck, wrecking, C-1 (AC)	16																														
Tank, light, M5	16																														
Trk-tractor, 5-6T, w/ semi-tlr., ponton	14																														
Motor carriage, M8	16																														
H-10 loading (AASHO)																															
Truck, 6-T, cargo	18																														
Crane, trk-mtd., w/crane atchmnts. tlr.	15																														
Truck, 4-T, w/155-mm How., carr. M1	16																														
Tank, light, 18-T	18																														
Truck, 6-T, bridge	19																														
Truck, 2½-T, w/8-T tlr.	17																														
Tank, medium, M2A1	21																														
Truck, 7½-T, cargo & prime mover	21																														
Tractor, D-8, w/dozer	22																														
Truck, 4-T, cargo, w/8-T tlr.	20																														
Truck, 6-T, w/3-in. AA, M2A2	22																														
Truck, 6-T, w/90-mm AA, M1	23																														
Motor carriage, M7	24																														
Trk-tractor, 6-T, w/semi-tlr., wrecking, C-2	26																														
H-15 loading																															
Motor carriage, M12	27																														
Motor carriage, M10	29																														
Trk-tractor, 7½-T, w/semi-tlr., fuel serv., F-1 (AC)	26																														
Truck, 7½-T, w/155-mm gun, carr. M2 & M3	28																														
Trk-tractor, 5-6T, w/ 20-T semi-tlr.	32																														
Truck, 6-T, w/16-T tlr	31																														
Tank, medium, M3	33																														
Tank, medium, M4	34																														
H-20 loading																															
Truck, 7½-T, w/8-in. gun, carr. M2, transp M1	34																														
Truck, 6-T, w/20-T tlr.	37																														
Truck, 7½-T, w/20-T tlr.	39																														
Tank, assault, T-14	46																														
Tank, heavy, M6	60																														

LEGEND
- SAFE
- CAUTION
- UNSAFE
- SS
- DS
- TS
- DD
- TD
- W/ = WITH

TABLE I

Vehicle capacity of fixed steel panel bridge, Bailey type.

6

TABLE II.—*Size of working parties*

Type of bridge	When equipment is stacked at site		When equipment must be unloaded from trucks	
	NCO's	EM	NCO's	EM
Single-single	4	33	8	55
Double-single	4	36	8	58
Triple-single	5	52	9	82
Double-double	5	64	9	106
Triple-double	6	90	10	138

For organization of working parties see paragraph 45.

TABLE IIIa.—*Estimated time for daylight construction*

Length in feet	Types of bridge				
	Single-single	Double-single	Triple-single	Double-double	Triple-double
40	1 hr. 15 min.	1 hr. 30 min.	——	——	——
60	1 hr. 30 min.	1 hr. 45 min.	2 hr.	——	——
80	——	2 hr.	2 hr. 30 min.	2 hr. 45 min.	——
100	——	2 hr. 15 min.	3 hr.	3 hr. 15 min.	4 hr.
120	——	2 hr. 30 min.	3 hr. 30 min.	3 hr. 45 min.	4 hr. 45 min.
140	——	——	4 hr.	4 hr. 15 min.	5 hr. 30 min.
160	——	——	——	4 hr. 45 min.	6 hr. 15 min.
180	——	——	——	——	7 hr.

TABLE IIIb.—*Estimated time for blackout construction*

Length in feet	Types of bridge				
	Single-single	Double-single	Triple-single	Double-double	Triple-double
40	1 hr. 45 min.	2 hr. 15 min.	——	——	——
60	2 hr.	2 hr. 30 min.	3 hr.	——	——
80	——	2 hr. 45 min.	3 hr. 30 min.	4 hr. 15 min.	——
100	——	3 hr.	4 hr.	4 hr. 45 min.	6 hr.
120	——	3 hr. 15 min.	4 hr. 30 min.	5 hr. 15 min.	6 hr. 45 min.
140	——	——	5 hr.	5 hr. 45 min.	7 hr. 30 min.
160	——	——	——	6 hr. 15 min.	8 hr. 15 min.
180	——	——	——	——	9 hr.

7

TABLE IV.—*List of equipment for one unit of fixed steel panel bridge, Bailey type*

Item	Quantity
Bar, carrying	84
Bearer, footwalk	72
Bearing	20
Bolt, bracing	480
Bolt, chord	120
Bolt, end post (spares)	4
Bolt, riband (curb)	214
Brace, sway	30
Chess	318
Clamp, transom	195
Davit	2
Extractor, pin	4
Footwalk	32
Frame, bracing	60
Jack, chord	8
Lever, panel	8
Link, launching nose	8
Panel, bridge	120
Pedestal, ramp	16
Picket, steel	40
Pin, panel	300
Pin, safety	315
Pin, sway brace (spares)	12
Plate, tie	60
Post, end, female	12
Post, end, male	12
Post, footwalk	72
Raker	45
Ramp, button	16
Ramp, plain	24
Riband (curb)	46
Roller, plain	12
Roller, rocking	12
Rope	
Lashing	15
Cordage	4
Shoe, jack	8
Stringer, button	30
Stringer, plain	45
Template, roller, plain	8
Template, roller, rocking	8
Timber, packing, 3 x 6 in. x 4 ft. 6 in.	12
Timber, grillage 6 x 6 in. x 4 ft. 6 in.	64
Transom	38
Wedge, hardwood, pair	8

Note: The words "panel bridge" found in the Engineer Supply Catalog nomenclature of the above items, have been omitted here for simplicity.

Item	Quantity
Hammer, hide faced, 4¼ lb. 2 in. face w/2 extra rawhide faces	20
Jack, ratchet-lever, automatic lowering double socket 15-ton	10
Sledge, blacksmith's, double-faced, 8-lb. w/handle	12
Spike, wire, type B, class 5, round, steel, flat head, ⅜ x 8 in. (lb.)	10
Wrench	
Reversable, ratchet	
1⅛ in. socket	15
1⅞ in. socket	15
Structural, offset, short tang	
1⅛ in.	15
1⅞ in.	15
Socket, offset 1⅛ in.	15

DESCRIPTION AND USE OF EQUIPMENT

---

7. PANEL.—The panel (fig. 6) is the basic member of the bridge. It is a truss section 10 feet long, 5 feet 1 inch high, and 7 inches wide. It weighs 600 pounds, and can be carried by six men using carrying bars.

The horizontal members of the panel are called chords. The bottom chords have four flat spaces with dowels on which the transoms fit. Also on the bottom chords are holes into which the ends of the sway brace fit. Both top and bottom chords have holes for the chord bolts which bolt the panels one above the other in building a double-story bridge. Additional holes in the top and ends of the panel accommodate bracing bolts that secure the rakers joining transoms and panels, and that also secure the bracing frames and tie plates joining panels of adjacent trusses.

Both chords of a panel have male lugs at one end and female lugs at the other. Panels are joined end to end by engaging these lugs and inserting panel pins in the hole in the lugs.

8. PANEL PIN.—The panel pin (fig. 9 (3)), 8 inches long and $1\frac{7}{8}$ inches in diameter, weighs 6 pounds. To make insertion easier it has a tapered end with a small hole for a safety pin to be inserted. Panel pins must be inserted with the groove on the head horizontal in order that the safety pins may be inserted.

9. TRANSOM.—The transom (fig. 7) is the principal cross member of the bridge. It is a 10-inch I-beam with a $4\frac{1}{2}$-inch flange, and is 18 feet long. It weighs 443 pounds, and is carried by six men using carrying bars inserted through holes in the web.

Figure 6.—Panel.

On the under side of the transom are six holes into which the dowels of the panels fit. On the upper side of the transom are five lugs which position the stringers, and near each end is an additional lug to which the raker, or side strut, is secured. Near each end are lugs on the web and on the flange of the *I*-beam to which the footwalk bearer is engaged.

Transoms support the floor system of the bridge. They rest on the lower chords of the panels, and are held in position by transom clamps. Transoms are at 5-foot spacing for loads under 40 tons. For heavier loads two transoms are used in the middle and one at each end of each panel (fig. 2). The bridge never must be jacked up under transoms as the transom clamps will fail.

10. TRANSOM CLAMP.—The transom clamp (fig. 9(1)) is a hinged, screw-type clamp, 13½ inches high and 8 inches across the top, weighing 7 pounds. It is placed on top of the transom and engaged in a slot in the panel. It is tightened by a vice-handled-type screw.

11. RAKER.—The raker (fig. 8(1)) is a 3-inch *I*-beam with a 1½-inch flange, 3 feet 6 inches long, and weighing 26 pounds. A raker connects the panel and transom at one end of each panel, by bracing bolts. It secures panels against a tendency to overturn. An additional raker is used at each end of the bridge. When there is more than one panel, the raker is connected to the inner one. At each end of the raker is a hollow dowel which engages in a hole in the panel and a hole in the transom.

12. BRACING FRAME.—A bracing frame (fig. 8(2)) is a rectangular frame, 4 feet 2 inches by 1 foot 8 inches, with a hollow conical dowel in each corner. It weighs 43 pounds.

Figure 7.—Transom.

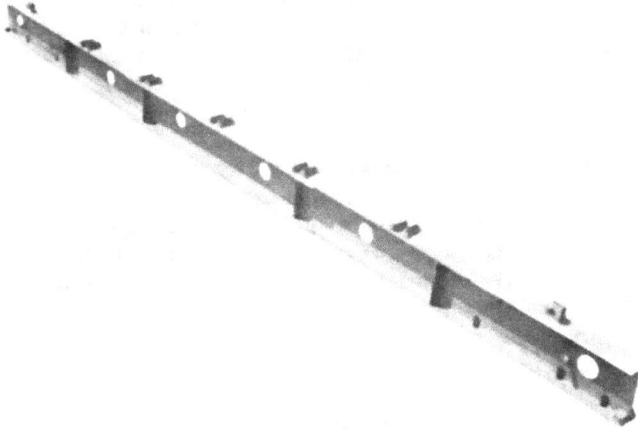

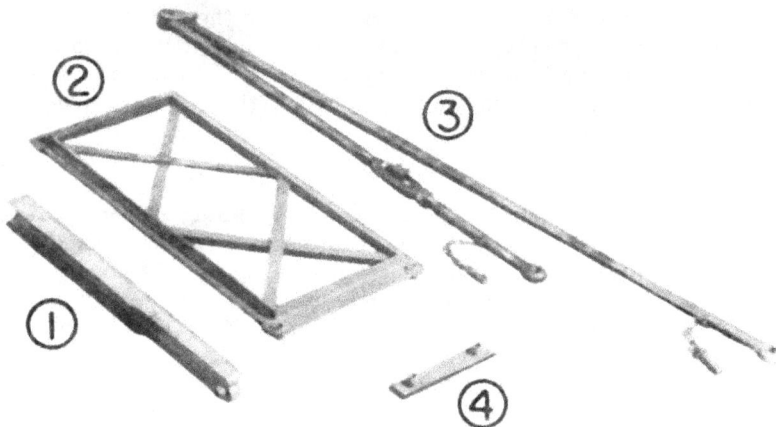

Figure 8.—(1) Raker; (2) bracing frame; (3) sway brace; (4) tie plate.

The bracing frame is used to brace double-and triple-truss bridges. In a double-single bridge, bracing bolts attach it horizontally to the top of the center of each pair of panels. In a double-double bridge, in addition to a horizontal bracing frame on the upper pair of panels, another is attached vertically, above the raker, on the ends of each upper pair of panels. On triple-truss bridges the bracing frame is similarly used, but only on the two inner trusses.

13. SWAY BRACE.—The sway brace (fig. 8(3)) is a 1-inch steel rod, hinged at the center, and adjusted by a turnbuckle. It weighs 63 pounds. At each end is an eye, through which a pin on a chain is inserted to secure it to the panel. The sway brace is given the proper tension by inserting the tail of an erection wrench in the turnbuckle and screwing up tight. The lock nut is then screwed up against the turnbuckle. Two sway braces are required in each bay of bridge.

14. TIE PLATE.—A tie plate (fig. 8(4)) is a piece of flat steel $2\frac{1}{2}$ by $\frac{3}{8}$ by 10 inches, and weighs 3 pounds. It has a hollow, conical dowel at each end.

The tie plate is used only in triple-truss bridges; it secures the second truss to the third truss, using the unoccupied raker holes in the panels at each joint and at the ends of the bridge. In triple-double bridges it is used in both lower and upper stories, but only at the joints of the upper story, the ends having bracing frames instead. A tie plate is used on one side of each joint and a bracing frame on the other.

15. BRACING BOLT.—A bracing bolt (fig. 9(5)) is ¾ inch in diameter, 3½ inches long, and weighs about 1 pound. A special lug on its head prevents rotation when the bolt is tightened.

It is used to: (1) Attach rakers to transoms and panels, (2) attach bracing frames to panels, (3) attach tie plates to panels. It is inserted into the hollow dowels of the braces to draw parts into proper alinement.

Figure 9.—(1) Transom clamp; (2) chord bolt; (3) panel pin with safety pin; (4) riband bolt; (5) bracing bolt.

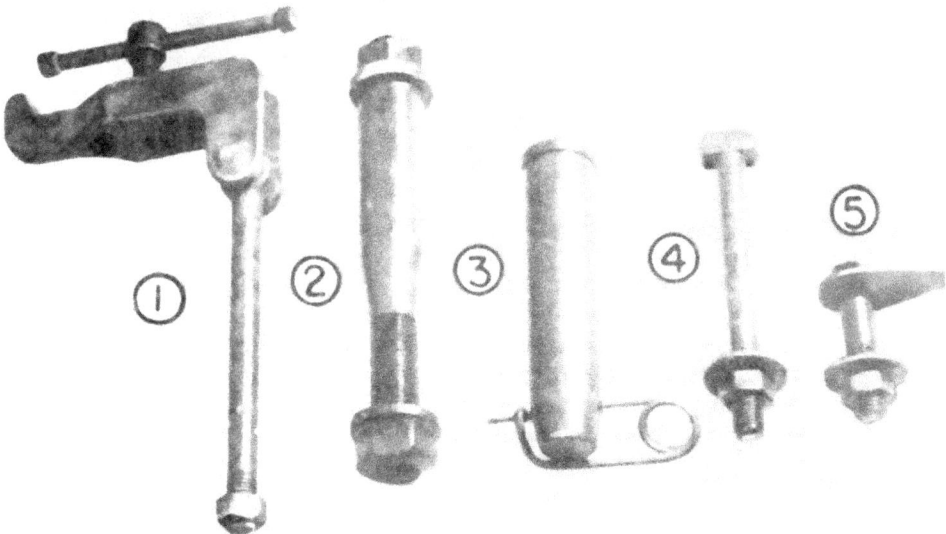

16. CHORD BOLT.—A chord bolt (fig. 9(2)) is 1¼ inches in diameter, 12¼ inches long, and weighs 8 pounds. It has a taper through half its length to assist in drawing the panels into alinement.

Chord bolts join upper and lower panels of double-story bridges. Two bolts per panel pass upward through holes in the chords of the panels, and are tightened with nuts on the lower chord of the upper story.

17. STRINGERS.—Each stringer consists of three 4-inch steel *I*-beams, 10 feet long, joined by welded braces. There is a

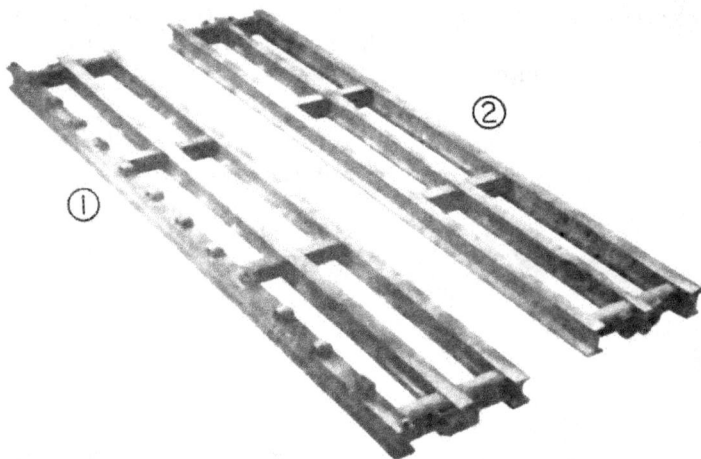

Figure 10.—(1) Button stringer; (2) plain stringer.

plain stringer (fig. 10(2)), which weighs 260 pounds, and a button stringer (fig. 10(1)) which weighs 270 pounds. They are identical except that the latter has 12 buttons which hold the chess in place. Four buttons have holes for the riband bolts.

Stringers carry the roadway of the bridge. Each bay has five stringers, three plain stringers in the middle, and a button stringer on each side.

18. CHESS.—A chess (fig. 11(1)) is 8-¾ by 2 inches by 11 feet 11 inches, is made of wood, and weighs 54 pounds.

Chess form the road surface. Each bay contains 13 chess, which lie across the stringers and are held in place by the buttons on the stringers. Chess are held down by ribands.

19. RIBAND (CURB).—A riband (curb) (fig. 11(2)) is a trapezoidal timber, 4 by 5½ by 5½ inches in cross section. It is 10 feet long, and weighs 77 pounds.

Ribands are used as curbing. They are fastened to the button stringers by four riband bolts.

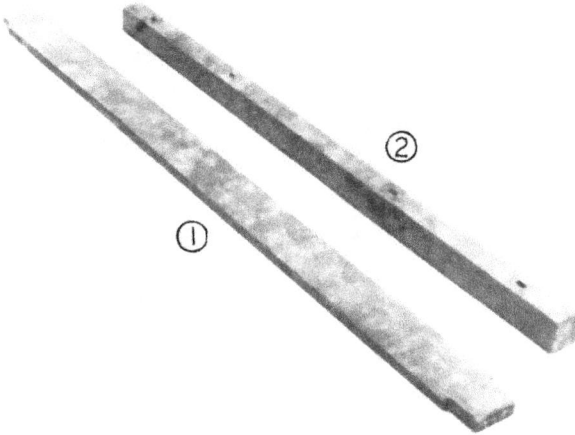

Figure 11.—(1) Chess; (2) riband.

20. RIBAND BOLT.—A riband bolt (fig. 9(4)) is T-headed, ¾ inch in diameter, 9 inches long, and weighs 1½ pounds. The thread is burred over at the top so the nut will not come off.

The riband bolt fastens the riband to button stringers and ramps. It is inserted head first through the slot in the riband until the head enters the button, then turned 90°, and the nut tightened with a wrench.

15

Figure 12.—(1) Female end post; (2) male end post.

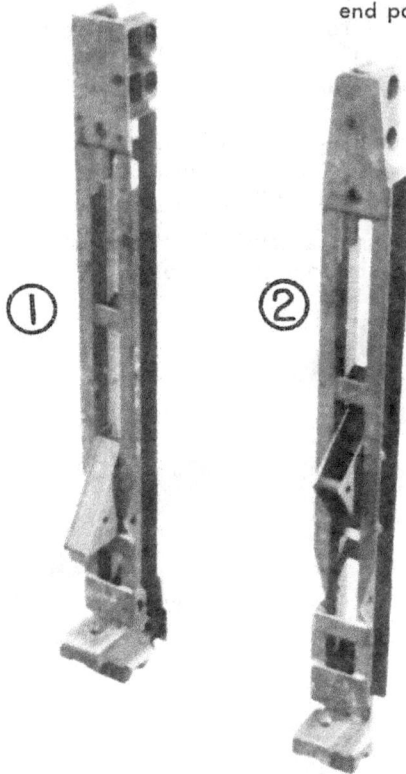

21. END POSTS.—End posts are of two types, weighing 120 pounds and 125 pounds respectively. They are columns, 5 feet high, of plates welded together. The two types have male (fig. 12(2)) and female lugs (fig. 12(1)) respectively, which are secured to the end panels of the bridge by panel pins fitting through holes in the lugs. End posts have a step, supporting a transom outside the panel at one end of the bridge. In jacking the bridge the jack is placed under the step. The lower end of the end post has a half-round bearing block which fits over the bearing.

16

End posts are used on both ends of the bridge to take the vertical shear. In double-story bridges they are placed only on the lower story.

22. BEARING.— A bearing (fig. 13) is a welded steel assembly weighing 99 pounds and containing a round bar upon which, when the bridge is completed, the bearing blocks of the end posts rest. During construction, the bearing block of the rocking roller rests upon it. The bar is divided into three parts by two intermediate sections that act as stiffeners.

The bearing spreads the load of the bridge over the ground or over grillage timber. When grillage timber is used, the bearing is held in place by spikes driven through holes in its base plate. If grillage is not used, pickets are driven through the holes into the ground. In a single-truss bridge, the bearing blocks of the end posts rest upon the middle part of the bearing; in a double-truss bridge two bearings are used, on each side of the bridge, each truss resting on the middle part (fig. 14); on each side of a triple-truss bridge, the inner truss rests on the middle part of one bearing, and the outer trusses rest on the outer parts of a second bearing.

23. RAMPS.—Ramps are similar to stringers, consisting of three 5-inch steel *I*-beams, 10 feet long, joined by welded braces. The lower surface of the ramp tapers upward near the ends. There is a plain ramp (fig. 15(2)) weighing 350 pounds, and a button type (fig. 15(1)) weighing 356 pounds. They are identical except the latter has 12 buttons which hold the chess in place. Four of the buttons have holes for the riband bolts.

Figure 13
Bearing

17

Figure 14.—End posts resting on bearings in double-truss bridge.

Three plain and two button ramps form continuations of the stringers leading from the bridge to the banks. If the slope is too steep, double-length ramps are used, supported at the joint by a transom held in place by four ramp pedestals. For loads over 40 tons the ramps are supported at their mid-points by packing timber. The transverse braces of the ramps are located and secured by the lugs on the transoms at the ends of the bridge.

24. RAMP PEDESTAL.—Ramp pedestals (fig. 16) are built-up, welded steel assemblies weighing 93 pounds. They keep the transoms which support double-length ramps from overturning. They are held in place by spikes or pickets driven through holes in their base plates. They are placed as shown in figures 59 and 60.

25. FOOTWALK.—A footwalk is 2 feet 6 inches wide and 10 feet long, and weighs 106 pounds. It is built of wooden duck-boards. Supported on the footwalk bearers, they are laid along the outer sides of the bridge for use by foot troops.

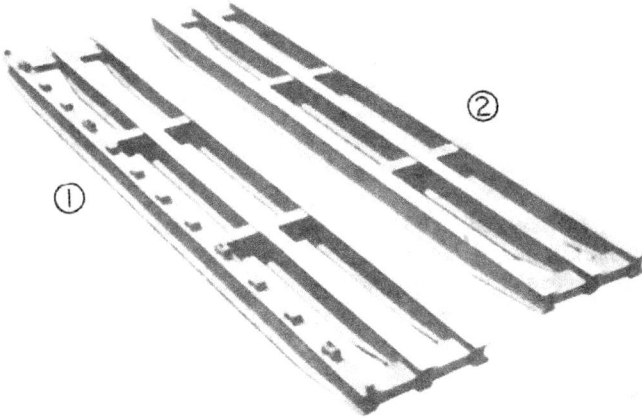

Figure 15.—(1) Button ramp; (2) Plain ramp.

Figure 16.—Ramp pedestal.

26. FOOTWALK BEARER.—A footwalk bearer is a built-up beam of pressed steel, 4 feet long and weighing 22 pounds. Bearers are attached to all transoms except reinforcing transoms, fitting over and under special lugs welded to the web near the ends of the transom. On top of the bearers are lugs between which the footwalk fits. At the end of the bearer is a socket to hold the footwalk post.

27. FOOTWALK POST.—Into every footwalk bearer is fitted a footwalk post 4 feet high and weighing 9 pounds. Hand ropes are threaded through two eyes on the top of each post and secured to holdfasts on the banks, or to the end footwalk posts.

28. ROCKING ROLLER.—The rocking roller, weighing 202 pounds (fig. 17) consists of three rollers housed in a balanced arm which fits over the bearing and is free to rock on it. Two side rollers on the flange on each side of the rocking roller frame act as guides for the trusses. The side rollers may be removed from the flanges by removing split pins from spindles underneath the flange; they then remain loosely attached to the frame by a chain.

By distributing the load when the bridge is launched, the rollers protect the bottom chord of the trusses when the bridge is rolled out over the bank seat (fig. 18). One pair of rollers is required for single-truss bridges, and two pairs for double-truss (fig. 19) and triple-truss bridges. On triple-truss bridges

Figure 17.—Rocking roller on bearing.

Figure 18.—Truss on rocking roller at bank seat.

the rollers are placed only under the inner and second trusses, and the outer side rollers of the outer rocking rollers are removed. One pair of rocking rollers is normally required on the far bank; two pair are used if the skeleton launching nose is double-truss in any part.

Figure 19.—Two rocking rollers used on double-truss bridge.

29. PLAIN ROLLER.—The plain roller (fig. 20) is 2 feet wide, and consists of a welded housing containing a single roller split in two. They weigh 105 pounds. The maximum allowable load on one roller is 6 tons.

During launching, plain rollers are placed every 25 feet behind the rocking rollers and at other required places except at the bank seats. Trusses of single-truss bridges may be carried on either half of the roller. Trusses of double- and triple-truss bridges are carried on both halves.

30. JACK.—The jack (fig. 21) used to lift the bridge on and off the rocking rollers, is a mechanical lifting jack of the type normally used in rigging and construction work. It has a 15-ton capacity and a lifting range of 15 inches. When the weight is carried on its toe, its capacity is only $7\frac{1}{2}$ tons. It weighs 112 pounds.

31. JACK SHOE.—The jack shoe (fig. 21) is a welded assembly used under the jack and designed to fit the bearing. In jacking under the step of the end posts the bearing can be put in place readily without moving the jack shoe.

Figure 20.—Plain roller.

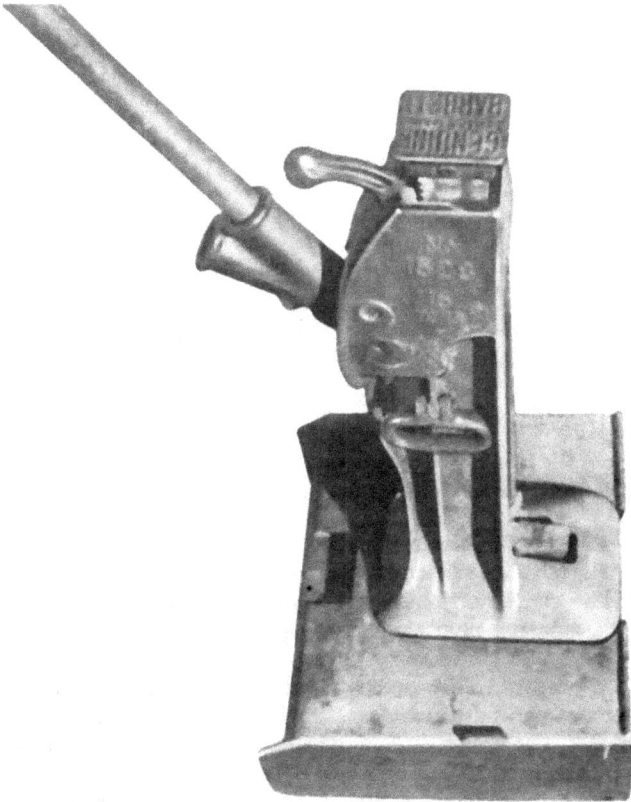

Figure 21.—Jack and jack shoe.

32. WRENCHES.—Four types of wrenches are provided to tighten or remove bolts and sway braces:

    a. 1⅛-inch offset socket wrench (fig. 22(1)).

    b. 1⅛-inch offset structural wrench (fig. 22(2)).

    c. 1⅞-inch offset structural wrench (fig. 22(3)).

    d. Reversible ratchet wrench handle (fig. 22(4)) with two removable socket heads—1⅞-inch (fig. 22(5)) and 1⅛-inch sizes (fig. 22(6)).

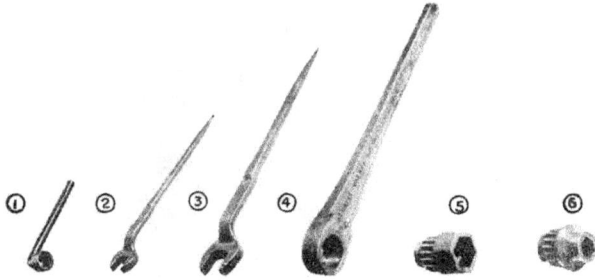

Figure 22.—(1) 1-⅛-inch socket wrench; (2) 1-⅛-inch structural wrench; (3) 1-⅞-inch structural wrench; (4) ratchet wrench handle; (5) I-⅞-inch ratchet wrench head; (6) 1-⅛-inch ratchet wrench head.

33. LEVER.—The lever (fig. 23(1)) which assists in erection of the second and third trusses after the first truss is in place over the gap, is a wooden bar 7 feet long weighing 45 pounds, with a fulcrum near the center and a lifting link at the end. The lifting link has a swiveling crosspiece which can be attached readily to the top of a panel by passing it beneath the upper chord, and turning it. The upper end of the link slides in a slot, the inner end of which is used when erecting the second truss and the outer end when erecting the third truss. The fulcrum always is placed on the top of the first truss. Two levers per panel are required, with two men operating each lever.

Figure 23.—(1) Lever; (2) carrying bar; (3) sledge; (4) picket.

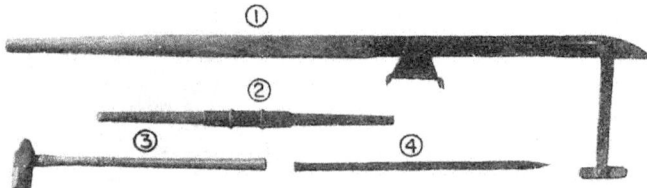

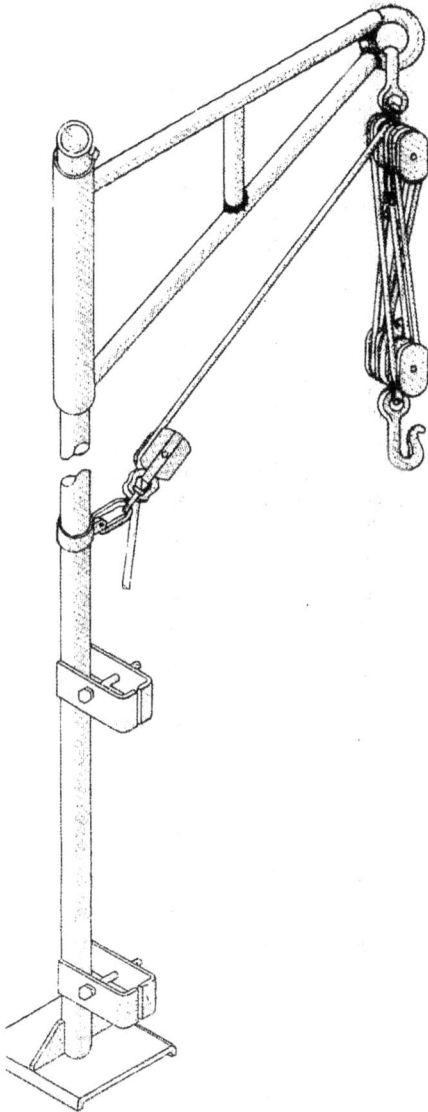

34. CARRYING BAR.—A wooden carrying bar (fig. 23(2)) 3 feet 6 inches long, reinforced by a steel band at the middle, is used to carry panels and transoms. It weighs 4 pounds.

35. DAVIT.—The davit (fig. 24) is 13 feet 9 inches in height. Complete with two ⅝-inch triple blocks, one ⅝-inch snatch block, and 90 feet of ⅝-inch rope, it weighs 227 pounds. It consists of a vertical tubular member with a short swiveling arm at the top. A bracket at the bottom of the tubular member positions the foot of the davit on top of the lower chord of a panel of the inner truss; another bracket, 5 feet above, positions the davit by a pin which passes through the bracket and engages with the chord bolt hole in the top chord of the panel. A triple-strand rope tackle is suspended from the arm, with the running end passing through a snatch block hooked into an eye just above the upper positioning bracket on the vertical member.

The davit is designed to lift and put into place the second and third trusses, after the bridge has been launched as a single-truss bridge.

Figure 24.—Davit.

25

36. CHORD JACK.—The chord jack (fig. 25) consists of two welded steel frames which fit on the top chord and engage with the plates which carry the holes for bracing-frame dowels. Each frame is held down by a *T*-headed bolt which passes up through the chord and the frame and is tightened by a nut on top of the frame. A knuckle-threaded screw assembly fits between the frames, and is operated by a ratchet lever to force them apart. The lever has a shackle at its end to which a rope may be attached to facilitate operation.

When adding a second story to a bridge already launched, work should start in the middle and proceed toward the ends. Upper jaws of the second-story panels will be found to be over-engaged, because of the natural sag of the bridge. Application of the chord jack spreads the jaws and allows the upper panel pins to be inserted.

37. PIN EXTRACTOR.—The pin extractor (fig. 26) assists in dismantling the bridge. After the pin has been partially driven out, and the recess under the head of the pin is exposed, the pin extractor grips this recess and forces out the pin by a levering action. It is particularly useful for dismantling the third truss of a triple-truss bridge, as the proximity of the second truss makes it impossible to drive out the pins with a hammer.

Figure 25.—Chord jack.

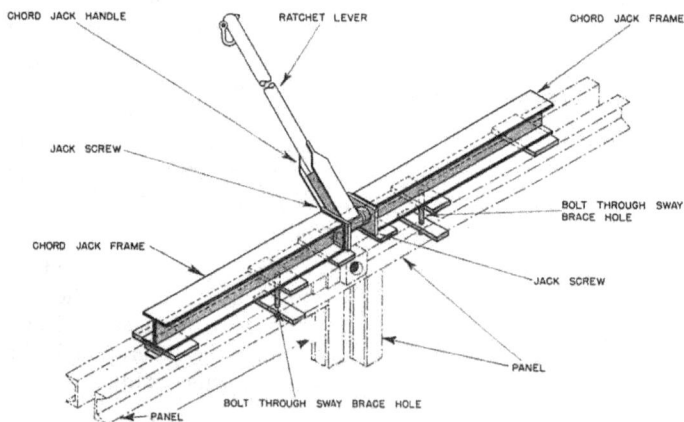

38. LAUNCHING-NOSE LINK.—The launching-nose link is about 10 inches long and 7 inches wide, and weighs 12 pounds. Links (fig. 27) consist of two steel frames, welded back to back, into which fit the lugs of two panels, joining them. The sides of the links have holes into which panel pins may be inserted. The links lie flush with the underside of the bottom chords of the panels, and have a false flange welded on the bottom edge so the bridge can be rolled out on launching rollers.

The launching-nose links overcome the sag occurring when the launching nose is cantilevered over the gap. Using one link on each truss increases the length of the bottom chords of the launching nose by 7½ inches, thus raising the end of the launching nose by 13½ inches for each bay ahead of the links. The links must not be inserted with more than four bays of the launching nose ahead of them; consequently, the maximum amount of lift that can be obtained from one pair of links is about 54 inches. If a greater lift is required, an additional pair of links can be used in one of the joints between the original pair and the end of the nose, its place depending on how much lift is required. (See table XIX.) The maximum lift obtainable is approximately 94½ inches.

Figure 26.—Pin extractor.

Figure 27.—Launching-nose link.

39. TEMPLATES.—Two types of templates are provided, one to locate the bearings for the rocking rollers and the other for the plain rollers.

a. The *rocking roller template,* (fig. 28(1)) weighing 100 pounds, consists of a timber base with timber strips on top which surround two spaces of sufficient size to accommodate the bearings on which the rocking rollers rest. At one end of the template are two angle cleats against which the braces of a stringer butt. This enables the stringer to act as a distance gage to position the templates, one on each side of the bridge (fig. 28(2)).

b. The *plain roller template,* (fig. 29(1)) weighing 30 pounds, consists of a smaller timber base with timber strips on three sides and a steel strip on the fourth side, to accommodate the base of a single plain roller. Two angle cleats at one end enable a stringer to be used as a gage in a manner similar to that described above, but only when one pair of plain rollers is used (fig. 29(2)). When two pairs are used, two plain roller templates may be used as a base, back to back; but in this case, a stringer cannot be used as a gage, and a tape must be used for measuring (fig. 29(3)).

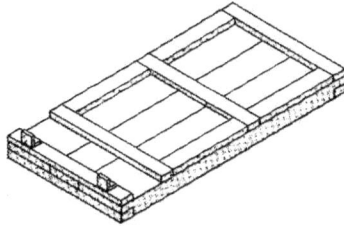

① ROCKING ROLLER TEMPLATE

SIDE ROLLERS REMOVED FOR TRIPLE TRUSS

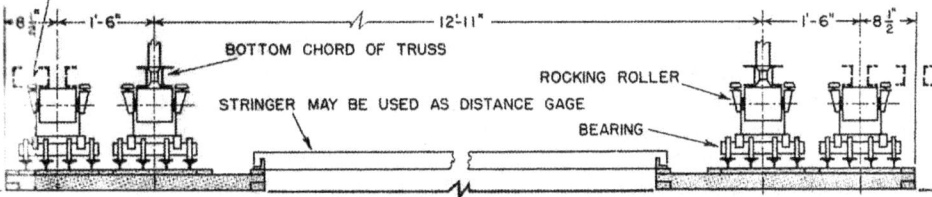

$8\frac{1}{2}''$ — 1'-6" — 12'-11" — 1'-6" — $8\frac{1}{2}''$

BOTTOM CHORD OF TRUSS

ROCKING ROLLER

STRINGER MAY BE USED AS DISTANCE GAGE

BEARING

② LAY-OUT OF ROCKING-ROLLER TEMPLATE

Figure 28.—(1) Rocking roller template; (2) lay-out of rocking roller
template.

Figure 29.—(1) Plain roller template; (2) lay-out of plain roller template for single or double truss; (3) lay-out of template
for double or triple truss.

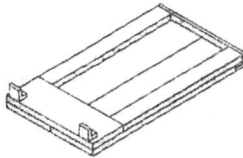

① PLAIN ROLLER TEMPLATE

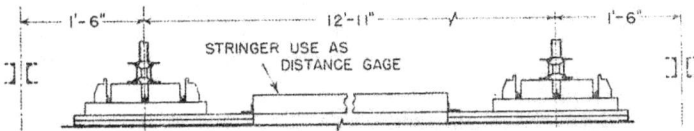

1'-6" — 12'-11" — 1'-6"

STRINGER USE AS
DISTANCE GAGE

② PLAIN ROLLER — SINGLE OR DOUBLE TRUSS

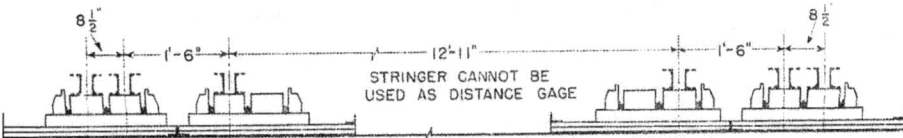

$8\frac{1}{2}''$ — 1'-6" — 12'-11" — 1'-6" — $8\frac{1}{2}''$

STRINGER CANNOT BE
USED AS DISTANCE GAGE

③ PLAIN ROLLER — DOUBLE OR TRIPLE TRUSS

TRANSPORTATION

40. MEANS OF TRANSPORTATION.—The panel bridge set is carried on twenty-one 2½-ton trucks drawing two-wheel pole-type trailers. All parts and erection tools can be manhandled on and off the trucks and trailers. Small parts should be kept in suitable boxes except when actually in use. Containers for this purpose are being designed. Until these containers are issued, parts may be kept in the boxes in which they are received, or others can be improvised.

41. TYPES OF LOADS.—The kinds of loads and the number of each in a bridge set are as follows:

Load	Number carried on trucks	Number carried on trailers
Interior	15	15
Ramp	4	4
Tool-and-footwalk	2	2
Total	21	21

a. Interior load (fig. 30).— The truck alone carries enough equipment for one bay of double-single bridge; with the trailer, it carries enough for one bay of double-double bridge. The composition of an interior load is given in table V.

b. Ramp load (fig. 30).— Enough equipment for a ramp bay is carried on one truck; a truck with trailer carries two ramp bays. The composition of a ramp load is given in table VI.

c. Tool-and-footwalk load.— The tools and erection equipment are carried on the trucks; the footwalk equipment is carried on the trailers. The composition of a tool-and-footwalk load is shown in table VII.

TABLE V.—*Interior Load*

Item	Unit weight (lb.)	Truck Quantity	Truck Total weight (lb.)	Trailer Quantity	Trailer Total weight (lb.)
Panels	600	4	2,400	4	2,400
Transoms	443	2	886		
Panel pins	6	10	60	10	60
Plain stringers	260	3	780		
Button stringers	270	2	540		
Sway braces	63	2	126		
Bracing frames	43	2	86	2	86
Rakers	26	3	78		
Tie plates	3	4	12		
Transom clamps	7	13	91		
Chess	54	14	756		
Ribands	77	2	154		
Chord bolts	8	4	32	4	32
Riband bolts	1½	10	15		
Bracing bolts	1	20	20	12	12
Safety pins	1/10	21	2		
Reversible ratchet					
wrench-1⅛-in. socket	15	1	15		
1⅞-in. socket	15	1	15		
1⅞-in. structural					
wrench	7	1	7		
1⅛-in. structural					
wrench	2	1	2		
1⅛-in. socket offset					
wrench	2	1	2		
Rope lashing	5	1	5		
Carrying bars	4	4	16		
Total			6,100 lbs.		2,590 lbs.

42. NUMBER OF LOADS REQUIRED.—The loads as given in tables V, VI, and VII allow for small spare parts, and include enough equipment for a 150-foot double-double bridge or for one 80-foot and one 70-foot triple-single bridge. The following loads are required for panel bridges:

31

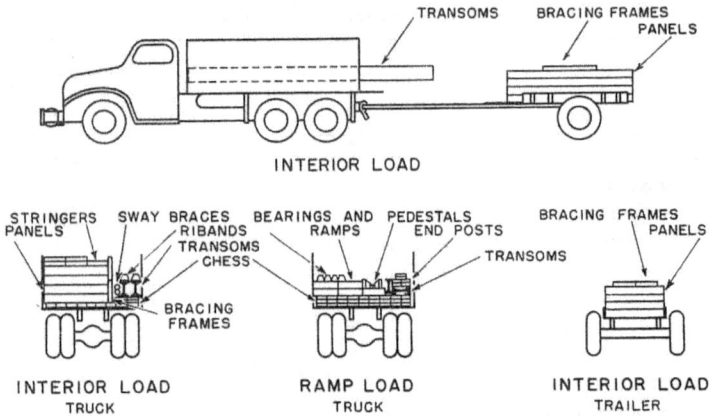

Figure 30.—Interior and ramp loads for fixed steel panel bridge, Bailey type.

Kind of bridge	Interior loads	Ramp loads	Tool-and-footwalk loads
Single-single Double-single	1 truck per 10 feet of bridge	1 truck per ramp bay or 1 truck w/trailer per 2 ramp bays	2 trucks w/trailers
Triple-single Double-double	1 truck w/trailer per 10 feet of bridge		

TABLE VI.—*Ramp load*

Item	Unit weight (lb.)	Truck Quantity	Truck Total weight (lb.)	Trailer Quantity	Trailer Total weight (lb.)
Male end posts	120	3	360		
Female end posts	125	3	375		
Plain ramps	350	3	1,050	3	1,050
Button ramps	356	2	712	2	712
Transoms	443	2	886		
Bearings	98	5	490		
Chess	54	27	1,458		
Ribands	77	4	308		
Riband bolts	1½	16	24		
Ramp pedestals	93	4	372		
End-post bolt	1	1	1		
Sway brace pins	1	3	3		
Total			6,039 lbs.		1,762 lbs.

TABLE VII.—*Tool-and-footwalk load*

Item	Unit weight (lb.)	Quantity	Total weight (lb.)
Truck			
Pickets	12	20	240
Rocking rollers	202	6	1,212
Plain rollers	105	6	630
Rawhide hammers	4	10	40
Jacks	112	5	560
Jack shoes	64	4	256
Carrying bars	4	12	48
Levers	45	4	180
Davit with tackle	227	1	227
Chord jacks	86	4	344
Pin extractors	15	2	30
Launching-nose links	12	4	48
Rocking roller templates	100	4	400
Plain roller templates	30	4	120
Grillage timbers	38	32	1,216
Packing timbers	18	6	108
Sledge hammers	12	6	72
Wedges	5	4	20
Total			5,751 lbs.
Trailer			
Footwalk bearers	22	36	792
Footwalks	106	16	1,696
Footwalk handposts	9	36	324
Cordage	20	2	40
Total			2,852 lbs.

SELECTION AND PREPARATION OF BRIDGE SITES

43. SITE REQUIREMENT.—The following site requirements are desirable:

a. Road over which bridge equipment can be moved to bridge site.

b. Approaches requiring little preparation.

c. Firm and stable banks.

d. Banks of about same height.

e. Cleared space long enough for assembly of bridge, and wide enough for unloading and stacking the parts and erection tools. The approach road often affords such a space.

f. Near an area suitable for a motor park and affording cover and concealment.

g. Surrounding terrain affording natural protection for security details.

44. PREPARATION OF SITE.—*a.* *General.*— A bulldozer may be employed to clear out and grade the bridge site and to construct a turn-around and motor park. Lacking a bulldozer, the work generally can be done with the pioneer tools carried on the equipment trucks.

b. *Turn-around.*— If existing roads do not afford a turn-around, generally one must be constructed to prevent traffic congestion.

c. *Motor park.*— A motor park is necessary for parking the equipment trucks.

d. *Clearing bridge site.*— The area around the site must be clear, and level enough for assembling the bridge and stacking equipment. It should be at least as long as the width of the gap to be spanned, and about 150 feet wide.

e. *Lay-out of equipment.*— For details of a suggested lay-out of equipment at bridge site see figure 31.

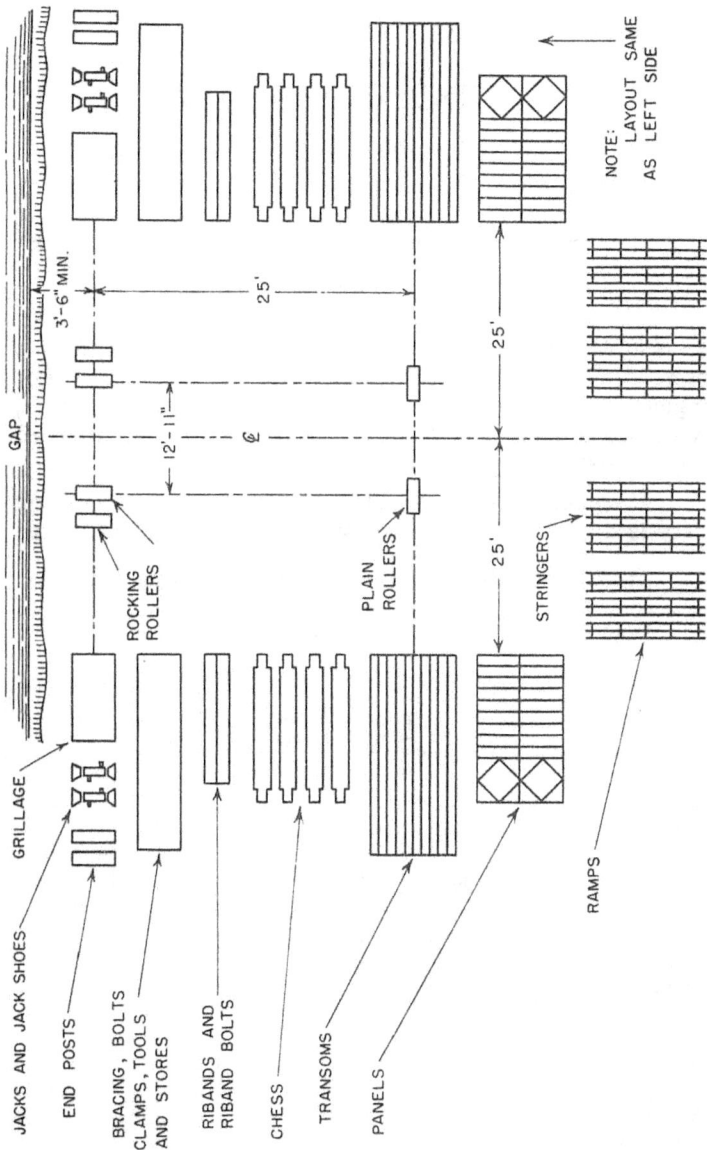

Figure 31.—Lay-out of equipment at bridge site.

WORKING PARTY

45. ORGANIZATION OF WORKING PARTY.—*a.* The bridge is erected under an officer, who supervises all operations. For single-story bridges over 100 feet long, or double-story bridges, an additional officer is necessary. A senior noncommissioned officer assists in supervision.

b. If equipment must be unloaded from trucks the working party is divided into construction and unloading details, **the**

TABLE VIII.—*Organization of construction details*

Detail	Number of NCO's and EM				
	Type of bridge				
	Single-single	Double-single	Triple-single	Double-double	Triple-double
Panel	1 - 14	1 - 14	2 - 28	2 - 32	3 - 50
Carrying	(12)	(12)	(24)	(28)	(44)
Pin	(2)	(2)	(4)	(4)	(6)
Transom	1 - 7	1 - 8*	1 - 8*	1 - 8*	1 - 8*
Carrying	(6)	(6)	(6)	(6)	(6)
Clamp	(1)	(2)	(2)	(2)	(2)
Bracing	1 - 4	1 - 6	1 - 8	1 - 12	1 - 20
Sway brace	(2)	(2)	(2)	(2)	(2)
Raker	(2)	(2)	(2)	(2)	(2)
Bracing frame	—	(2)	(2)	(4)	(4)
Chord bolt	—	—	—	(4)	(8)
Tie plate	—	—	(2)	—	(4)
Decking	1 - 8	1 - 8	1 - 8	1 - 12	1 - 12
Stringer	(4)	(4)	(4)	(8)	(8)
Chess and riband	(4)	(4)	(4)	(4)	(4)
Total	4 - 33	4 - 36	5 - 52	5 - 64	6 - 90

*Add 1 NCO and 8 EM when transoms are doubled in bridges designed to carry loads over 40 tons.

latter under a senior noncommissioned officer who also dispatches trucks to the motor park. Organization of construction details is shown in table VIII, and of unloading details in table IX.

c. If an unloading detail is necessary, it is organized as shown in table IX.

d. The organization of construction details given in table VIII does not provide for launching the bridge. When the bridge is to be pushed forward, all other work stops and the entire working party launches it. If too heavy to be pushed across by manpower, a truck or a bulldozer may be used.

e. Normally, the construction details can start work as soon as the site is prepared, without waiting for completion of unloading.

f. Additional men must be provided for construction of approaches, traffic control, and security patrols, if these tasks are to be carried on simultaneously with construction of the bridge.

TABLE IX.—*Organization of unloading detail*

	Number of NCO's and EM				
	Type of bridge				
	Single-single	Double-single	Triple-single	Double-double	Triple-double
Unloading detail	4 - 22	4 - 22	4 - 30	4 - 42	4 - 48
Panel	(1)-(6)	(1)-(6)	(1)-(12)	(1)-(12)	(1)-(18)
Transom	(1)-(6)	(1)-(6)	(1)-(6)	(1)-(12)	(1)-(12)
Bracing	(1)-(2)	(1)-(2)	(1)-(4)	(1)-(6)	(1)-(6)
Deck	(1)-(8)	(1)-(8)	(1)-(8)	(1)-(12)	(1)-(12)
Total construction details (Table VIII)	4 - 33	4 - 36	5 - 52	5 - 64	6 - 90
Total working party	8 - 55	8 - 58	9 - 82	9 -106	10-138

Figure 32.—Panel detail carrying panel.

46. DUTIES OF CONSTRUCTION DETAILS.—The duties of construction details are as follows:

a. The panel detail.— (1) Carries (fig. 32), places, and pins together panels in launching nose and bridge.

(2) As soon as all panels are in place, divides into two crews. One crew crosses to far bank and begins dismantling launching nose. The other carries necessary parts to far bank for completion of end of bridge and installation of ramp.

(3) Reforms as a single detail and completes dismantling of launching nose.

(4) Installs far-bank end posts.

(5) Jacks down far end of bridge.

(6) Installs far-bank ramp, and places its chess and ribands.

b. The transom detail.— (1) Carries, places (fig. 33), and clamps down transoms (fig. 34).

(2) Removes plain rollers on near bank.

(3) Installs end posts on near bank.

(4) Assists decking detail in jacking down near end of bridge.

(5) Installs near-bank ramp and assists decking detail in placing chess and ribands on it.

c. The decking detail.— (1) Lays out rocking and plain rollers on near bank.

38

(2) Assists panel detail in starting construction of launching nose.

(3) Carries rollers to far bank and lays them there.

(4) Returns to near bank and lays stringers, chess, and ribands on bridge (fig. 35).

(5) Jacks down near end of bridge.

(6) Lays chess and ribands on near-bank ramp.

d. *The bracing detail.*— Obtains, installs, and adjusts:

(1) Sway braces (fig. 36).

(2) Rakers (fig. 37).

(3) Bracing frames (fig. 38)—on double- and triple-truss bridges only.

(4) Chord bolts (fig. 39)—double-story bridges only.

(5) Tie plates—on triple-truss bridges only.

Figure 33.—Transom detail installing panel in place.

47. PARTS REQUIRED PER BAY OF BRIDGE.—The tools and equipment necessary to construct panel bridges vary with site conditions, lengths of spans, and number of trusses. The parts required per bay of bridge are constant, and may be determined from table X.

Figure 34.—Transom clamp in place.

Figure 35.—Decking detail laying stringers, chess, and ribands.

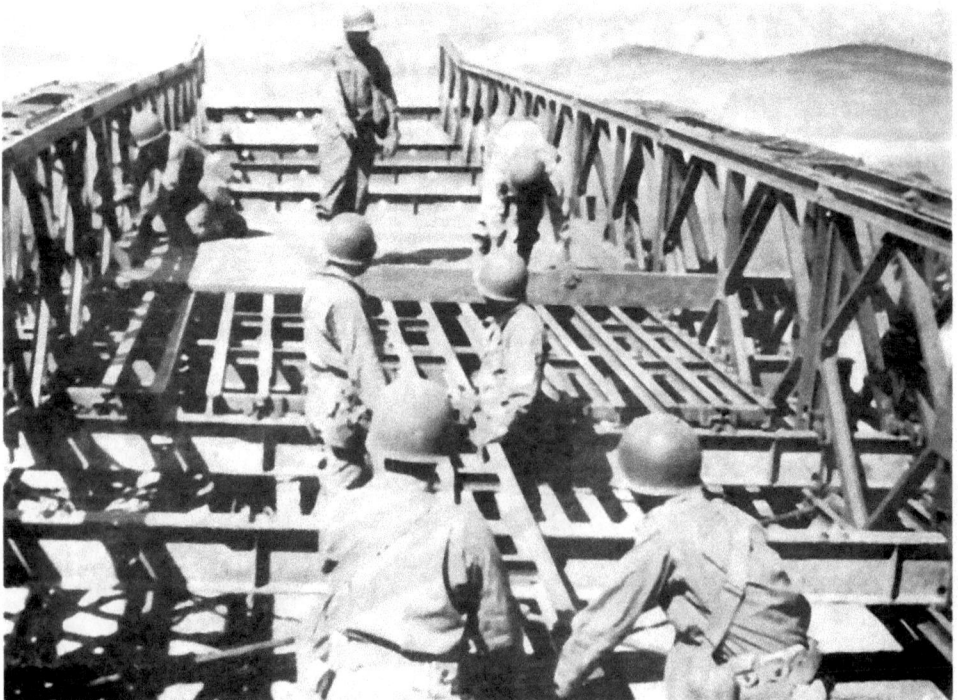

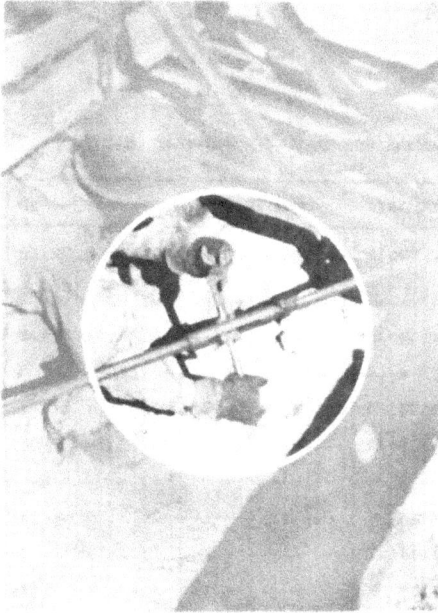

Figure 36.—Sway brace being tightened.

Figure 37.—Raker being bolted.

Figure 38.—Bracing frame being bolted.

Figure 39.—Chord bolt being tightened.

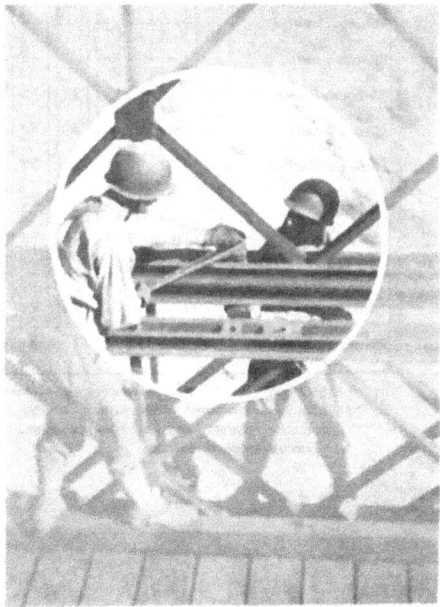

TABLE X.—*Number of parts required per bay for various bridges.*

PART	END BAY (No. 1)*					INTERIOR BAY					RAMP BAYS**					END BAY (No. 2)*				
	SS	DS	TS	DD	TD	SS	DS	TS	DD	TD	SS	DS	TS	DD	TD	SS	DS	TS	DD	TD
Bearings	2	4	4	4	4	—	—	—	—	—	—	—	—	—	—	2	4	4	4	4
Bracing bolts	4	12	16	20	28	4	12	16	20	28	—	—	—	—	—	8	16	24	32	40
Bracing frames	—	2	2	4	4	—	2	2	4	4	—	—	—	—	—	—	2	2	6	6
Chess	13	13	13	13	13	13	13	13	13	13	26	26	26	26	26	13	13	13	13	13
Chord bolts	—	—	—	8	12	—	—	—	8	12	—	—	—	—	—	—	—	—	8	12
End posts, male	2	4	6	4	6	—	—	—	—	—	—	—	—	—	—	—	—	—	—	—
End posts, female	—	—	—	—	—	—	—	—	—	—	—	—	—	—	—	2	4	6	4	6
Footwalks	2	2	2	2	2	2	2	2	2	2	—	—	—	—	—	2	2	2	2	2
Footwalk bearers	4	4	4	4	4	4	4	4	4	4	—	—	—	—	—	6	6	6	6	6
Footwalk posts	4	4	4	4	4	4	4	4	4	4	—	—	—	—	—	6	6	6	6	6
Panels	2	4	6	8	12	2	4	6	8	12	—	—	—	—	—	2	4	6	8	12
Panel pins with safety pins	4	8	12	12	18	4	8	12	16	24	—	—	—	—	—	8	16	24	28	42
Rakers	2	2	2	2	2	2	2	2	2	2	—	—	—	—	—	4	4	4	4	4
Ramp pedestals	—	—	—	—	—	—	—	—	—	—	4	4	4	4	4	—	—	—	—	—
Ramps, button	—	—	—	—	—	—	—	—	—	—	4	4	4	4	4	—	—	—	—	—
Ramps, plain	—	—	—	—	—	—	—	—	—	—	6	6	6	6	6	—	—	—	—	—
Ribands (curbs)	2	2	2	2	2	2	2	2	2	2	4	4	4	4	4	2	2	2	2	2
Riband bolts	8	8	8	8	8	8	8	8	8	8	16	16	16	16	16	8	8	8	8	8
Stringers, button	2	2	2	2	2	2	2	2	2	2	—	—	—	—	—	2	2	2	2	2
Stringers, plain	3	3	3	3	3	3	3	3	3	3	—	—	—	—	—	3	3	3	3	3
Sway braces	2	2	2	2	2	2	2	2	2	2	—	—	—	—	—	2	2	2	2	2
Tie plates	—	—	2	—	4	—	—	2	—	4	—	—	—	—	—	—	—	4	—	4
Transom***	2	2	2	2	2	2	2	2	2	2	1	1	1	1	1	3	3	3	3	3
Transom clamps***	4	8	12	8	12	4	8	12	8	12	—	—	—	—	—	4	8	12	8	12

* End bay (No. 1) assumed to be bay without transom on end post.
** Ramp bays assumed to be double length.
*** For loads over 40 tons, add two transoms in end and interior bays, and increase transom clamps proportionately.

CONSTRUCTION

48. METHODS OF LAUNCHING.—*a.* The panel bridge can be launched by any one of the following methods:
    (1) Cantilever, employing a skeleton launching nose.
    (2) Cantilever, employing a tail or counterweight.
    (3) Derrick or gin pole.
    (4) Rolling out on falsework.
    (5) Floating out on pontons over waterways.
  b. At most sites, the easiest and quickest procedure is the cantilever method given in (1), combined when necessary with a counterweight as given in (2). The bridge is assembled on rollers on one bank and pushed across the gap, with sufficient weight kept behind the rollers to maintain balance and prevent the bridge from tipping into the gap. A skeleton framework, called a launching nose, placed on the front of the bridge assists its travel into position on the far side. With this method it is essential that the bridge be properly balanced during the launching and that the launching nose be constructed with a definite number and type of bays as explained below.

49. SIZE OF BRIDGE.—The type of bridge constructed is governed by the load capacity required by the tactical situation. Its length is some multiple of 10 feet, the minimum being the width of the gap to be spanned plus 3 feet at each bank seat for the bearings. If the ground is soft, or the banks unstable, bearings should be placed farther back.

50. PLACING LAUNCHING ROLLERS.—As soon as the gap has been measured, the launching site cleared, and the site for stacking equipment marked off, the launching rollers are placed. The area for the grillage—if needed—on which the bearings and rocking rollers rest on the near bank then is carefully leveled. The amount of grillage used is determined

Figure 40.—Rocking rollers in position on bank seat.

by the heights of the near- and far-banks and by the nature of the soil.

Next, rocking rollers (two for single-truss and four for double- and triple-truss construction) resting on bearings, are placed on each side of the gap 12 feet and 11 inches apart, measuring from center-to-center of the rollers (fig. 40.). The centers of the rollers are placed at least 2 feet 6 inches, and normally about 3 feet 6 inches, toward the gap from the location of the bearings for the end posts of the completed bridge. Therefore, if the final bridge spans 150 feet, the center-to-center distance between rocking rollers must not exceed 145 feet. This permits the bridge to be jacked down from the rollers onto the bearings. Consequently, in tables XI through XVII the distance between rocking rollers is shown as 5 feet less than the span of the completed bridge.

Templates are provided to assist in locating the rollers (fig. 28), each template accomodating a second pair of rollers when required. Templates remain in position under the bearings during launching. The rocking rollers must be level, and perpendicular to the axis of the bridge. A carpenter's level on the stringer which locates the templates is a means of checking the level of the bearings under the rocking rollers. Normally, the rocking roller template may be used on the ground; but on soft soil, grillage should be used beneath it.

Two plain rollers—four for double or triple trusses—are placed every 25 feet behind the rocking rollers. The plain roller template is used to position them, and they are spaced as shown in figure 29. Usually the template provides sufficient bearing surface for the plain rollers, but in soft soil, grillage sometimes is necessary for footings.

The landing rollers on the far bank may be plain rollers if the load to be carried on each roller is not more than 6 tons. For heavier loads, rocking rollers are used. Either type of roller is placed at least 2 feet 6 inches toward the gap from the location of the bearings on which the far-bank end posts rest. Templates, with a stringer as a distance gage, also may be used here to position the rollers.

51. SKELETON LAUNCHING NOSE.—*a. General.*—The launching nose (fig. 41) consists of panels, transoms, rakers, sway braces, and when necessary, launching-nose links. It does not have stringers or decking. The first bay at the tip of the nose has two transoms; the other bays have one. Rakers are used *on every transom,* and sway braces *in the second bay from the tip and in every bay* thereafter. Footwalks are not built on the skeleton nose itself.

b. Composition of launching nose.—The number and type of bays used in the nose depends upon the span and the type of construction of the bridge. The composition of the launching nose for the various combinations of span and bridge construction is given in tables XI through XVII. *These tables must be followed exactly with respect to the composition of the launching nose.*

Figure 41.—Completed launching nose.

SPAN (FEET)	LAUNCHING WEIGHT(TONS)	SAG (INCHES)
30	9	3
40	12	4
50	15	$5\frac{1}{2}$
60	18	8

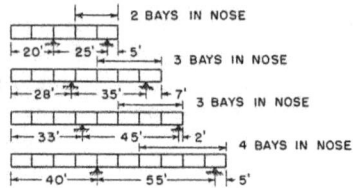

2 BAYS IN NOSE
|← 20' →|← 25' →|← 5'

3 BAYS IN NOSE
|← 28' →|← 35' →|← 7'

3 BAYS IN NOSE
|← 33' →|← 45' →|← 2'

4 BAYS IN NOSE
|← 40' →|← 55' →|← 5'

TABLE XI.—*Combinations for launching single-single bridges.*

NOTES:

1.   All launching assemblies shown are at point of balance at rocking rollers on near bank.

2.   Launching rollers must not be nearer than 2 feet 6 inches to final position of end posts as shown:

|← →|← 2'-6" MIN.

3.   Sag indicated is *approximate* sag at tip of launching nose, just before reaching far bank. These figures aid in determining number and position of launching-nose links required for launching nose to clear far bank.

4.   One pair of rocking rollers is required on both near and far banks.

46

SPAN (FEET)	LAUNCHING WEIGHT (TONS)	SAG (INCHES)
40	16	$2\frac{1}{2}$
50	20	$4\frac{1}{2}$
60	25	$7\frac{1}{2}$
70	25	11
80	29	15
90	33	20
100	36	25
110	40	31
120	44	38

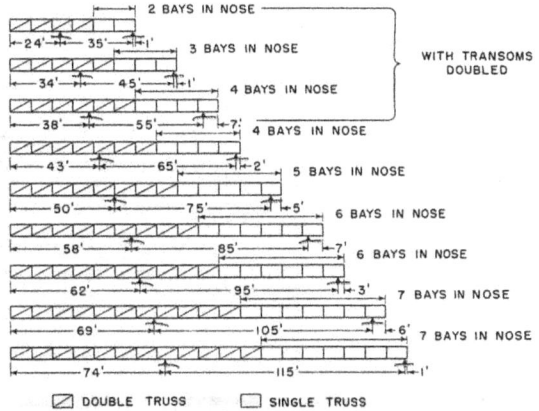

2 BAYS IN NOSE
24'—35'—1'
3 BAYS IN NOSE   WITH TRANSOMS DOUBLED
34'—45'—1'
4 BAYS IN NOSE
38'—55'—7'
4 BAYS IN NOSE
43'—65'—2'
5 BAYS IN NOSE
50'—75'—5'
6 BAYS IN NOSE
58'—85'—7'
6 BAYS IN NOSE
62'—95'—3'
7 BAYS IN NOSE
69'—105'—6'
7 BAYS IN NOSE
74'—115'—1'

☑ DOUBLE TRUSS   ☐ SINGLE TRUSS

TABLE XII.—*Combinations for launching double-single bridges.*

NOTES:

1. All launching assemblies shown are at point of balance at rocking rollers on near bank.

2. Launching rollers must not be nearer than 2 feet 6 inches to final position of end posts as shown:

2'-6" MIN.

3. Sag indicated is *approximate* sag at tip of launching nose, just before reaching far bank. These figures aid in determining number and position of launching-nose links required for launching nose to clear far bank.

4. Two pairs of rocking rollers are required on near bank, and one pair on far bank.

47

SPAN (FEET)	LAUNCHING WEIGHT (TONS)	SAG (INCHES)
40	18	2
50	23	$3\frac{1}{2}$
60	28	$5\frac{1}{2}$
70	33	$8\frac{1}{2}$
80	38	12
90	42	17
100	42	22
110	47	27
120	51	34
130	55	40
140	59	47

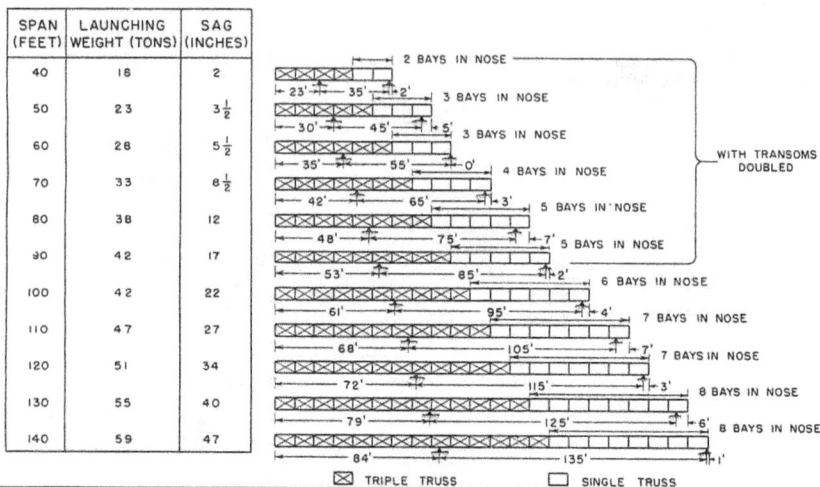

2 BAYS IN NOSE · 23' · 35' · 2' · 3 BAYS IN NOSE · 30' · 45' · 5' · 3 BAYS IN NOSE · 35' · 55' · 0' · 4 BAYS IN NOSE · 42' · 65' · 3' · 5 BAYS IN NOSE · 48' · 75' · 7' · 5 BAYS IN NOSE · 53' · 85' · 2' · 6 BAYS IN NOSE · 61' · 95' · 4' · 7 BAYS IN NOSE · 68' · 105' · 7' · 7 BAYS IN NOSE · 72' · 115' · 3' · 8 BAYS IN NOSE · 79' · 125' · 6' · 8 BAYS IN NOSE · 84' · 135' · 1' · WITH TRANSOMS DOUBLED · ⊠ TRIPLE TRUSS · □ SINGLE TRUSS

TABLE XIII.—*Combinations for launching double-double bridges. (Complete before launching)*

NOTES:

1. All launching assemblies shown are at point of balance at rocking rollers on near bank.

2. Launching rollers must not be nearer than 2 feet 6 inches to final position of end posts as shown:

2'-6" MIN.

3. Sag indicated is *approximate* sag at tip of launching nose, just before reaching far bank. These figures aid in determining number and position of launching-nose links required for launching nose to clear far bank.

4. Two pairs of rocking rollers are required on near bank, and one pair on far bank.

SPAN (FEET)	LAUNCHING WEIGHT (TONS)	SAG (INCHES)
80	43	12
90	49	14
100	53	17
110	59	21
120	58	25
130	63	30
140	69	37
150	74	43
160	79	50

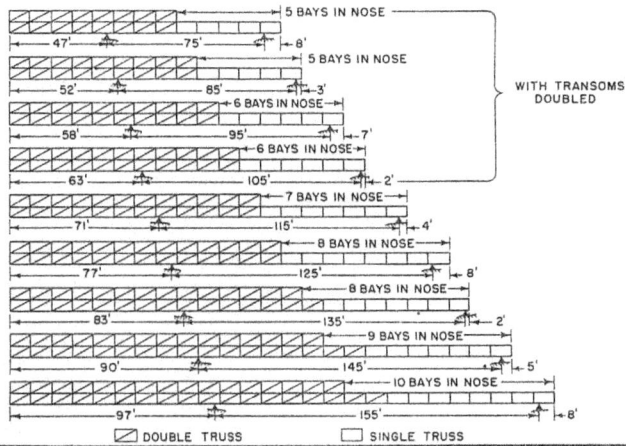

5 BAYS IN NOSE — 47' — 75' — 8'
5 BAYS IN NOSE — 52' — 85' — 3'
6 BAYS IN NOSE — 58' — 95' — 7'
6 BAYS IN NOSE — 63' — 105' — 2'
7 BAYS IN NOSE — 71' — 115' — 4'
8 BAYS IN NOSE — 77' — 125' — 8'
8 BAYS IN NOSE — 83' — 135' — 2'
9 BAYS IN NOSE — 90' — 145' — 5'
10 BAYS IN NOSE — 97' — 155' — 8'

WITH TRANSOMS DOUBLED

DOUBLE TRUSS   SINGLE TRUSS

TABLE XIV.—*Combinations for launching double-double bridges.*
(Incomplete before launching)

NOTES:

1. All launching assemblies shown are at point of balance at rocking rollers on near bank.

2. Launching rollers must not be nearer than 2 feet 6 inches to final position of end posts as shown:

2'-6" MIN.

3. Sag indicated is *approximate* sag at tip of launching nose, just before reaching far bank. These figures aid in determining number and position of launching-nose links required for launching nose to clear far bank.

4. Two pairs of rocking rollers are required on near bank, and one pair on far bank. However, when any part of launching nose consists of double-truss construction, two pairs of rocking rollers are required on far bank.

49

SPAN (FEET)	LAUNCHING WEIGHT (TONS)	SAG (INCHES)
80	39	10
90	44	14
100	49	19
110	54	24
120	54	30
130	58	37
140	63	44
150	67	52
160	72	59

Diagram annotations:

- 4 BAYS IN NOSE — 44' — 75' — 1'
- 5 BAYS IN NOSE — 50' — 85' — 5' — WITH TRANSOM DOUBLED
- 5 BAYS IN NOSE — 55' — 95' — 0'
- 6 BAYS IN NOSE — 62' — 105' — 3'
- 7 BAYS IN NOSE — 68' — 115' — 7'
- 7 BAYS IN NOSE — 73' — 125' — 2'
- 8 BAYS IN NOSE — 79' — 135' — 6'
- 8 BAYS IN NOSE — 84' — 145' — 1'
- 9 BAYS IN NOSE — 90' — 155' — 5'

DOUBLE TRUSS   SINGLE TRUSS

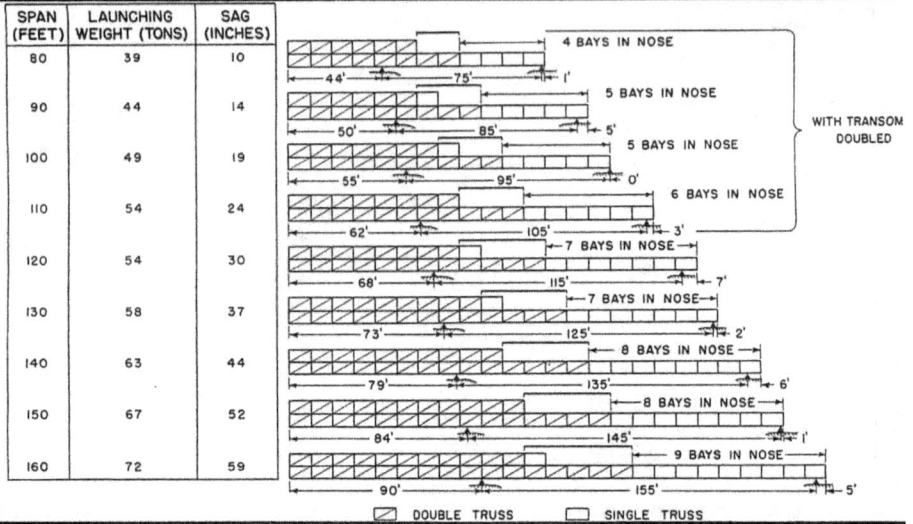

TABLE XV.—*Combinations for launching triple-single bridges.*

NOTES:

1. All launching assemblies shown are at point of balance at rocking rollers on near bank.

2. Launching rollers must not be nearer than 2 feet 6 inches to final position of end posts as shown:

2'-6" MIN.

3. Sag indicated is *approximate* sag at tip of launching nose, just before reaching far bank. These figures aid in determining number and position of launching-nose links required for the launching nose to clear far bank.

4. Two pairs of rocking rollers are required on near bank, and one pair on far bank. However, when any part of launching nose consists of double-truss construction, two pairs of rocking rollers are required on far bank.

5. Panels of launching nose are used to complete portions of bridge marked thus:

SPAN (FEET)	LAUNCHING WEIGHT (TONS)	SAG (INCHES)
100	62	15
110	69	20
120	75	25
130	80	31
140	81	36
150	81	42
160	86	49
170	92	55
180	97	61

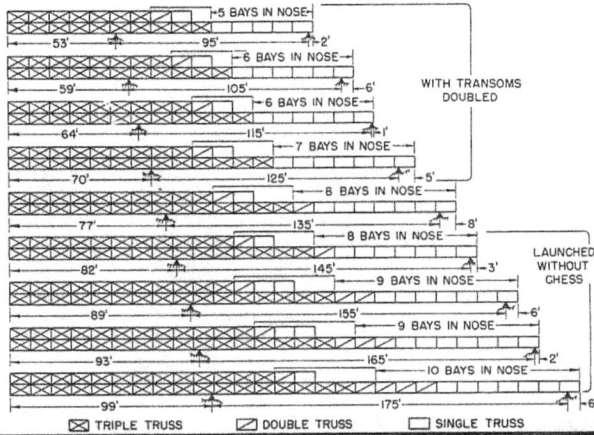

Diagram labels (right of table):

- 5 BAYS IN NOSE — 53', 95', 2' — WITH TRANSOMS DOUBLED
- 6 BAYS IN NOSE — 59', 105', 6'
- 6 BAYS IN NOSE — 64', 115', 1'
- 7 BAYS IN NOSE — 70', 125', 5'
- 8 BAYS IN NOSE — 77', 135', 8'
- 8 BAYS IN NOSE — 82', 145', 3' — LAUNCHED WITHOUT CHESS
- 9 BAYS IN NOSE — 89', 155', 6'
- 9 BAYS IN NOSE — 93', 165', 2'
- 10 BAYS IN NOSE — 99', 175', 6'

⊠ TRIPLE TRUSS   ⊡ DOUBLE TRUSS   ▭ SINGLE TRUSS

TABLE XVI.—*Combinations for launching triple-double bridges.*
(Incomplete before launching)

NOTES:

1. All launching assemblies shown are at point of balance at rocking rollers on near bank.

2. Launching rollers must not be nearer than 2 feet 6 inches to final position of end posts as shown:

2'-6" MIN.

3. Sag indicated is *approximate* sag at tip of launching nose, just before reaching far bank. These figures aid in determining number and position of launching-nose links required for the launching nose to clear far bank.

4. Two pairs of rocking rollers are required on both near and far banks.

SPAN (FEET)	LAUNCHING WEIGHT (TONS)	SAG (INCHES)
100	66	15
110	74	20
120	81	23
130	88	28
140	87	33
150	89	39
160	95	45
170	101	50
180	107	55

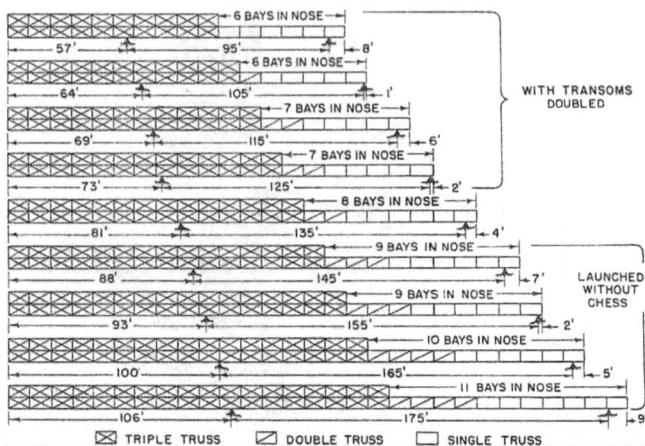

⊠ TRIPLE TRUSS   ▱ DOUBLE TRUSS   ▢ SINGLE TRUSS

TABLE XVII.—*Combinations for launching triple-double bridges.*
(Complete before launching)

NOTES:

1. All launching assemblies shown are at point of balance at rocking rollers on near bank.

2. Launching rollers must not be nearer than 2 feet 6 inches to final position of end posts as shown:

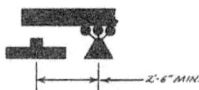

3. Sag indicated is *approximate* sag at tip of launching nose, just before reaching far bank. These figures aid in determining number and position of launching-nose links required for launching nose to clear far bank.

4. Two pairs of rocking rollers are required on near bank, and one pair on far bank.

5. Panels of launching nose are used to complete portions of bridge marked thus:

c. Use of launching-nose links.—The launching nose tends to sag as it is cantilevered over the gap. The *approximate* sag at the end of the nose just before it reaches the far bank is shown in tables XI to XVII. To overcome this sag and also to reach a far-shore bank seat which is higher than the near bank seat, launching-nose links are used (fig. 42) as described in paragraph 38. The vertical lift of the tip of the nose thus obtained is shown in table XIX. For an example of the use of launching-nose links, see appendix I.

When the far-bank seat is higher than the near-bank seat by a distance greater than the rise afforded by the links, timber grillage is used to raise the rollers on which the nose is launched. In the same situation launching-nose links may be omitted by building up the grillage sufficiently high beneath the rocking rollers. This, however, requires that the bridge be jacked down on the near bank a greater distance than normally is necessary, and entails use of block and tackle to prevent the bridge from sliding backwards.

Figure 42.—Upturned skeleton launching nose.

When the far-bank seat is lower than the near-bank seat, launching-nose links are unnecessary. However, it is necessary to use block and tackle on the near bank to prevent the bridge from running away when the center of gravity passes the rocking rollers.

d. Arrival of launching nose on far bank.—When the end of the nose reaches the far bank it is guided on the rollers previously placed. If space is restricted, or if the ground rises steeply away from the gap, the bridge is pushed forward one bay at a time and the skeleton panels are removed as they clear the rollers. The minimum length of level clear space required on the far bank is 12 to 15 feet. The panels removed from the launching nose may be used to strengthen the bridge, if required.

52. LAUNCHING OF THE BRIDGE.—As the skeleton nose is assembled it is placed on rollers. Where working space is restricted, part of the launching nose may be pushed out over the gap as it is assembled provided its center of gravity, or balance point, remains behind the rocking rollers. After the launching nose is complete, the bridge proper is assembled. With single story construction the bridge and nose may be pushed out over the gap after every two bays are constructed. With double story construction it is preferable to assemble the entire bridge before pushing it across the gap.

During launching, the entire bridge including launching nose must be counterbalanced so that the structure does not tip into the gap. The counterbalance is normally accomplished by adding sufficient bridge behind the near-shore rocking rollers *to act as a counterweight and keep the center of gravity, or point of balance, between the plain rollers and the rocking rollers. This condition must prevail until the launching nose reaches the rollers on the far bank.* This point is illustrated in tables XI to XVII which show the bridge and launching nose just spanning the gap; in this position the bridge is completely assembled and the center of gravity is slightly behind the near-side rocking rollers. As the bridge is pushed on across the gap from this position, the center of gravity passes the rocking rollers. The part of the bridge acting as a counterweight is no longer needed to maintain balance because there is now no danger of tipping into the gap; however, *it is needed to avoid excess stress in the launching nose until launching is completed.* Dismantling any of the bridge behind the rocking rollers will throw additional stress on the launching nose and on the part of the bridge which is across the gap; this may result in failure of the nose.

If lack of space or bridge material prohibit constructing the entire length of the bridge prior to launching, dead weight may be added to the tail of the bridge to act as a counterweight. The effect of the counterweight may be equal to the counterbalance effect of that part of the bridge which is shown behind the rocking rollers in tables XI through XVII. The amount of counterweight needed can be calculated by taking moments about the rocking rollers. As was explained above for normal launching, *the dead weight used as a counterweight must not be removed or allowed to cross in front of the rocking rollers until the entire launching nose has passed over the far-side rollers.* The procedure of using dead weights other than the bridge itself is not desirable and should be avoided where possible.

As mentioned above, in double story construction it is desirable to construct the entire bridge before launching. The launching nose combinations for double story bridges which are *completely assembled* before launching are shown in tables XIV and XVI. If there is a shortage of materials, certain panels may be omitted in the original assembly as shown in tables XV and XVII. These latter two tables apply only to bridges which are thus incomplete before launching. The launching nose for a double story bridge which is launched incomplete is not identical with the launching nose used for the same length of bridge launched complete. *Bays of the bridge shown as incomplete in tables XV and XVII should not be confused with bays of the launching nose.* The incomplete bays of the bridge proper require only additional panels to make them complete, while the launching nose bays do not have stringers, decking, or the required number of transoms. Panels from the disassembled launching nose are used to complete the bridge bays. *However, the launching nose itself is never converted into bridge by completing its panel and floor construction;* instead it is disassembled after it passes the rollers on the far side and parts so obtained may be added to the bridge proper.

Long triple-double bridges (over 150 feet) must be launched without chess in the floor system. This is indicated in tables XVI and XVII. The bridges are too heavy to handle if the chess are in place.

The following precautions must be observed during construction and launching:

a. DO NOT PLACE BENT OR DISTORTED PARTS IN ANY BRIDGE.

b. IN CONSTRUCTING THE LAUNCHING NOSE, PLACE RAKERS ON EVERY TRANSOM AND PLACE SWAY BRACES IN EVERY BAY EXCEPT THE FIRST.

c. CONSTRUCT THE LAUNCHING NOSE EXACTLY AS INDICATED IN TABLES XI THROUGH XVII.

d. DO NOT ATTEMPT TO CONVERT THE LAUNCHING NOSE INTO BRIDGE BY ADDING ADDITIONAL MATERIALS TO IT.

e. IN LAUNCHING BRIDGE OVER ROLLERS, KEEP THE CENTER OF GRAVITY BEHIND THE ROCKING ROLLERS UNTIL THE LAUNCHING NOSE REACHES THE FAR SIDE. THEREAFTER DO NOT DISMANTLE THE BRIDGE BEHIND THE NEAR-SIDE ROCKING ROLLERS OR REMOVE THE COUNTERWEIGHT UNTIL ALL OF THE LAUNCHING NOSE HAS CLEARED THE FAR-SIDE ROCKING ROLLERS.

Figure 43.—First bay of launching nose.

Figure 44.—Position of panel pins in completed bridge.

53. ASSEMBLY OF LAUNCHING NOSE AND BRIDGE.—After the grillage, bearings, and launching rollers have been positioned, proceed with assembling and launching the bridge as follows:

a. Place a panel over each inner rocking roller and a transom in position inside of the end upright, letting it rest on the lower chord of each panel. Clamp panels and transom together with transom clamps. Place a raker at each end of transom, and secure them with bracing bolts (fig. 43). To prevent each panel from rolling back, place a steel picket through bottom chord of panel and rocking roller.

b. Connect a second pair of panels to the first pair. Insert the panel pins with the points outward and with the grooves in the heads of the pins horizontal. See figure 44 and paragraph 8. In double- and triple-truss construction point the pins inward in the second and third trusses. Place a transom in front of the panel joints and clamp it to the panels.

57

In single-story construction, the transoms in the main part of the bridge should be behind the panel uprights, so stringers and decking can be placed on each bay as it is built. In double-story construction, the first transom should be behind the end upright, and the others in front of the uprights. The first bay of the bridge proper is left undecked until the bridge reaches its final position and the nose has been removed. This allows the decking detail to keep closer to the panel detail, and gives the men erecting the second story more room in which to work.

c. Place a pair of sway braces in the second bay (fig. 45).

d. If they are required, place launching-nose links in position between panels (fig. 46) as determined by erection conditions. See appendix I and figure 66 to determine the number of links and their position in the nose. If links are used, an additional pair of plain rollers should be placed halfway between the rocking and the plain rollers. These additional rollers should be 2 inches below the line of sight joining the other rollers. They prevent the launching-nose links from fouling the ground between the other rollers during launching. The additional rollers are removed as soon as the links have passed over the rocking rollers.

e. Continue adding panels (fig. 47) with transoms every 10 feet, sway braces in every bay, and rakers on every transom until sufficient skeleton nose has been built. For lengths of nose required for various spans see tables XI—XVII.

Figure 45.—Sway braces in place.

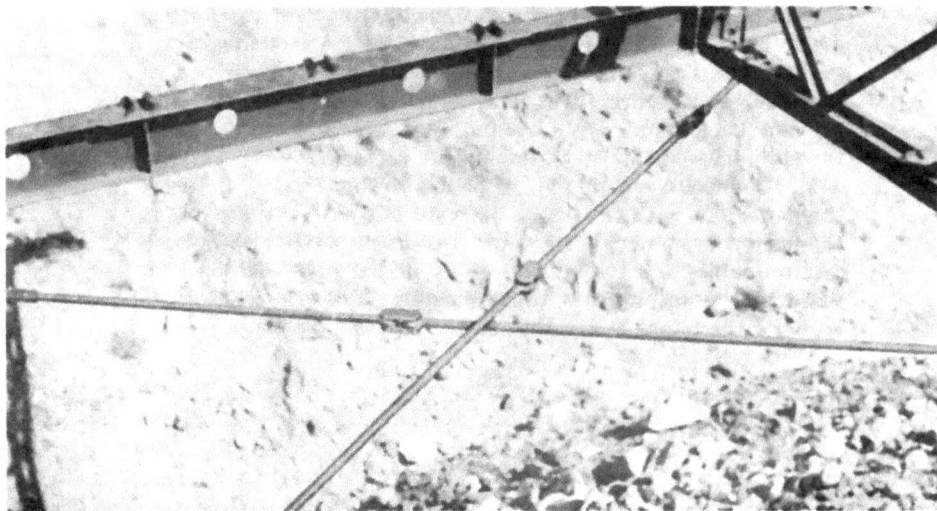

Figure 46.—Inserting panel pin to connect panel in launching nose to launching-nose links.

Figure 47.—Installing outer panel in first bay of bridge after completion of launching nose.

Figure 48.—Installing inner panel in third bay of bridge.

f. When construction of the nose is completed, continue construction by adding panels in every bay (fig. 48)—two in *SS,* four in *DS,* six in *TS* bridges, and so on. Add transoms every 5 feet on bridges for loads under 40 tons; for greater loads the number of transoms is doubled. All transoms in the bridge proper must be in front, or all in rear, of uprights of the panels; otherwise the decking will not fit. In double-truss bridges the transoms on either side of the middle upright of the panel are fastened with only one transom clamp each,

Figure 49.—Pushing launching nose over gap after completion of first two bays of bridge.

placed on opposite sides of alternate uprights.

g. After transoms are in position, place sway braces in each bay. Install rakers joining panels and those transoms which are at the panel joints. In double- and triple-truss construction fit one pair of bracing frames per bay, and in triple-truss construction also fit one pair of tie plates per bay.

All sway braces, transom clamps, bracing frames, rakers, and tie plates should be left loose until all parts of the bay except stringers and decking are fitted for the next bay ahead.

61

Figure 50.—Chess and ribands in place in partially completed bridge.

Then all these bracing devices should be tightened simultaneously.

h. Do not let a bridge get out of balance on the rollers; push it forward as panels are added to the rear (fig. 48). In launching a bridge up to 80 feet long the plain rollers should be about 25 feet behind the rocking rollers, provided the bridge is pushed forward to keep the center of balance behind the rocking rollers. When about half of a longer bridge has been built, insert another pair of rollers by pushing the bridge forward until it is balanced on the rocking rollers, then lifting the rear and placing the additional rollers. They should be about 25 feet behind the original pair of plain rollers and should be so seated that the tops of all rollers, when in position are in the same plane.

i. Add stringers bay-by-bay during assembly. Pass chess and ribands through the panel and place them; then bolt the ribands to the button stringers with riband bolts (fig. 50).

j. As bays of the bridge are completed, it is rolled out across the gap. When the forward end of the launching nose reaches the rollers on the far bank (fig. 51) it is guided onto the rollers (fig. 52) and is dismantled by the panel detail (fig. 53) bay-by-bay. The bridge may be pushed out a few bays at a time; but, until the launching nose reaches the far bank, the

Figure 51.—Launching nose approaching far bank; rocking rollers
in place.

center of gravity must be kept behind the rocking rollers (see
paragraph 52). Laying stringers and decking, and tightening
the various bracing devices, continue as the launching nose is
dismantled.

k. When the end of the bridge proper clears the rollers on
the far bank, attach the near-bank end posts, lay a transom
across their steps, and lay stringers and decking of the last
bay.

Simultaneously, attach the far-bank end posts and, if the
bridge is to carry loads over 40 tons, lay a transom across
their steps.

At the same time, place and level off the grillage on which
the bearings are to rest on both near and far banks. A transom can be used as a straight edge to line up the grillage.

l. When using bulldozers or trucks to launch bridges the
following precautions should be observed:

(1) Do not apply power directly to the end of a panel;
pressure should be against the end posts, or a transom supported by the end posts. When power is applied against a
transom it must be distributed across the entire length of
the transom.

(2) Adjust roller heights so the tail of the bridge is at
least 6 inches off the ground during the entire launching.

Figure 52.—Guiding launching nose onto rocking rollers.

(3)   Rig a snub line to control movement of the bridge.

(4)   If the bridge requires two trucks or bulldozers to move it, use one against each end post.

m.   Place the jack shoes on the grillage and the jacks upon the shoes, with the toes of the jack under the steps of the end posts (fig. 54).   At each corner of the bridge, use only the number of jacks required by the weight to be lifted (see table XVIII for weight of bays).   Jack up the ends of the bridge successively and remove the rocking rollers (fig. 55).   Place in position on the prepared grillage the bearings on which the rocking rollers were resting, and alongside place other bearings, if required, to take the other trusses.

Figure 53.—Dismantling launching nose.

Figure 54.—Jacks in place before jacking end of bridge off
rocking rollers.

Then lower the bridge in stages (figs. 56 and 57), placing
grillage under the bottom chords of the trusses in case the
bridge slips off the jacks. It does not matter which end of the
bridge is lowered first, but *the jacks must be operated in unison* and only on one end at a time.

n. If bridge is to carry loads of 28 tons or more, place
wedges under the midpoint of the end transoms (fig. 4). Next,
lay ramps (fig. 58). If the slope of the ramp is not over 1
foot in 10, only one bay of ramps is necessary (fig. 59). When
two ramp bays are used (fig. 60), they are supported at their

Figure 55.—Jacking up end of bridge before removal of
rocking rollers.

Figure 56.—Jacking down end of bridge on bearings.

TABLE XVIII.—*Weights per bay of various types of bridges.*

Type of bridge	Weight per bay in tons (normal transoms)	Weight per bay in tons (double transoms)
Single-single	2.39	2.85
Double-single	3.05	3.51
Triple-single	3.68	4.15
Double-double	4.34	4.80
Triple-double	5.60	6.07
Skeleton launching nose Single-single	.88	—
Skeleton launching nose Double-single	1.48	—

Figure 57.—End of bridge jacked down on bearings.

Figure 58.—First bay of ramp installed.

junction by a transom supported in turn by four ramp pedestals resting on the ground (fig. 61). Ramp bays are decked in the same way as bays of the bridge proper.

o. Footwalks should be erected before launching as it is awkward to place bearers and walks after bridge is in place. On a double-story bridge, erecting footwalks before launching provides a footing for the men placing the second story. However, when short of time or manpower, footwalks can be built after bridge is completed.

p. In double-truss construction, the inner truss is constructed first. The outer truss is added, with pins inserted pointing inward. The transoms then are inserted through the panels from the side, and the transom clamps are placed in position, but not tightened (see g above). Before the clamps

Figure 59.—Single-bay ramp completely decked.

CHAPTER 6

are finally tightened rakers and bracing frames are added in order.

Where it is not possible for the second truss to be added before the transoms are put in place—for example, in reinforcing a single-truss bridge—levers are used to assist in positioning the outer panels (see par. 34).

q. In triple-truss construction, successive bays should be added to the third truss, keeping one bay behind the other two. The third truss is connected to the second by tie plates bolted to the top raker holes in the uprights of the panels (fig. 62).

If a shortage of panels makes it necessary to use those of the skeleton launching nose to complete a double- or triple-truss bridge, erection should be started back of the first bays of the bridge proper. The panels omitted can be placed in position with davit (see par. 35) and levers after the bridge has been launched.

r. (1) In double-story construction, erection of the second story can start before or during launching, or after the gap has been spanned. However, the preferred method is to complete the second story before launching. Panels are carried out on the deck horizontally and from there lifted to the top of

Figure 60.—Second bay of ramp installed, chess laid on first bay.

Figure 61.—Close-up of ramp pedestal supporting transom.

the lower story. They are hoisted into vertical position by a panel-carrying crew assisted by two chord-bolt men (fig. 63). Next, the chord bolts are placed, but not tightened until the lower panel pins can be inserted without difficulty. The outer top truss is placed first. Add the second and third trusses successively, not both at once.

(2) To convert a bridge from one type to another, the chord jack is employed to insert the upper panel pins of the second story (see par. 36) in the following cases:

(*a*) When a second story is added to a completed bridge. The maximum spans that can be converted to double-story in this case are 120-foot double-truss, with or without footwalks; 140-foot triple truss, without footwalks, or 130-foot triple truss, with footwalks.

69

Figure 62.—Tie plate between second and third end panels in triple-single bridge.

(b)   When a double-truss, double-story bridge is converted to triple-truss, double-story.

(c)   When a double-story bridge is launched with a nose formed of panels which are to be used in the upper story launching. The upper story panels then may be fitted in position as soon as the launching is completed.

(3)   The following precautions must be observed in using chord jacks:

(a)   Jacking must commence at the middle of the bridge and move toward both ends simultaneously, except in case (2) (c) above, where it commences at the point nearest the middle of the bridge proper and moves toward the far bank.

(b)   Bracing frames should be fitted before jacking.

(c)   The ratchet lever of the chord jack must not be lengthened.

70

(*d*)   No more than three men should haul on the rat-chet lever at one time; if they cannot move the load, two jacks should be used simultaneously on adjacent trusses.

(4)   In double-story bridges, bracing frames are placed on the top chords, in the middle of each panel (fig. 64), and vertically on the rear face of the uprights at each panel joint (fig. 65).   They must be inserted before the chord bolts in the bay are tightened.   Connection between the second and third trusses is made with tie plates, as in single-story construction.

(5)   In constructing double-story bridges, the body of the bridge should be kept as far back on the bank as the site allows to make it easier to insert the second-story panel pins and chord bolts.   This can be done by using extra rollers as mentioned in *h* above.

54.   DISMANTLING BRIDGE.—In general, the bridge is dismantled by reversing the erection procedure.   First the ramps and ramp decking are removed; then the bridge is jacked up and rocking rollers on bearings are placed in the same positions as for launching.   Before the weight of the bridge is placed

Figure 63.—Placing second-story panel.

Figure 64.—Bracing frame in place in horizontal position in triple-truss bridge.

on the jacks, timber grillage is inserted beneath the bottom chords of the trusses, in case the jack slips. *The jacks must be operated in unison,* on one end at a time. The skeleton launching nose then is constructed of the same size required in launching. Next, the bridge is withdrawn over the gap by pushing from the far bank and pulling from the near bank by manpower or mechanical power. Trusses are dismantled by reversing the erection procedure. Before dismantling an individual bay, all bolts and braces are loosened and removed. As the bridge is dismantled, the individual parts are carried to their appropriate trucks and loaded as described in paragraph 41.

Figure 65.—Bracing frame placed vertically on ends of intermediate pair of panels on double-story bridge.

TRAFFIC CONTROL AND MAINTENANCE OF BRIDGE

---

55. TRAFFIC CONTROL.—Traffic over the bridge must be closely supervised to insure prompt passage of important tactical vehicles, to control passage of vehicles with weights exceeding the rated capacity of the bridge, to provide for two-way traffic, and to prevent damage to the bridge. For detailed discussion of traffic control see FM 5-10.

56. MAINTENANCE OF BRIDGE.—A bridge guard should be maintained, whose duties are to:

   a. Supervise traffic.

   b. Maintain bridge approaches.

   c. Provide local security.

   d. See that transom clamps and bracing, riband, and chord bolts are kept tight.

   e. See that panel pins do not work out, and that their safety pins are in position.

   f. Watch ground at bank seats for signs of settlement, and add grillage if necessary.

   g. Keep deck clear of stones and gravel.

   h. If bridge is semipermanent, lay a wearing surface of planking lengthwise of the bridge.

   i. Pour a small quantity of oil over each panel joint occassionally, if the bridge remains for a long period.

APPENDIX I

EXAMPLE OF THE USE OF LAUNCHING-NOSE LINKS
(See fig. 66)

Given:

Length and type of bridge	— 130-ft. double-double.
Distance between near- and far-bank rocking rollers	— 125 ft.
Ground level at far bank seat	— 36 in. above near bank seat.
Ground level at rear plain rollers	— 24 in. above near bank seat.
Distance from base of bearing to top of rocking rollers	— 16½ in.
Distance from base to top of plain rollers	— 7 in.

Figure 66.—Use of launching-nose links.

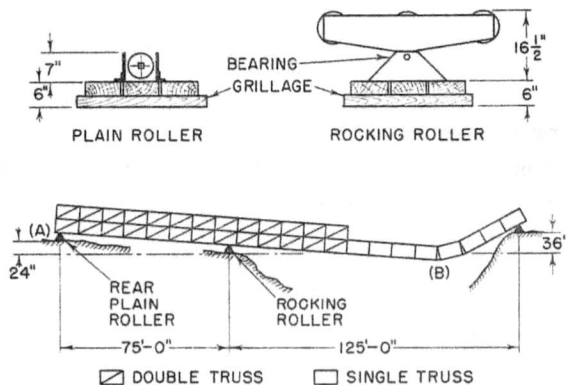

PLAIN ROLLER   ROCKING ROLLER

BEARING

GRILLAGE

$7"$ $6"$ $16\frac{1}{2}"$ $6"$

(A)   (B)

$24"$ $36"$

REAR PLAIN ROLLER   ROCKING ROLLER

75'–0"   125'–0"

▨ DOUBLE TRUSS   ☐ SINGLE TRUSS

74

Height of grillage beneath rocking rollers and bearing   — 6 in.

Height of grillage beneath plain
rollers   — 6 in.

Distance between rocking rollers
and rear plain rollers   — 75 ft.

Sag of launching nose for 130 ft.
double-double bridge (see
table XIV)   — 30 in.

Problem:

To find distance of links from tip of launching nose.

Solution:

(1) Difference in ground
level between rear plain
rollers and rocking rol-
lers   = 24 in.

Distance from ground
level to top of rear plain
rollers   = 7 in. + 6 in.   = 13 in.

Distance from ground
level to top of rocking
rollers   = 16½ in. + 6 in.   = 22½ in.

TABLE XIX.—*Use of launching-nose links*

Distance of links from tip of launching nose (feet)	Resulting vertical lift of tip of nose (inches)
One pair of links	
10	13½
20	27
30	40½
40	54
Two pairs	
10 and 40	67½
20 and 40	81
30 and 40	94½

Therefore, height of top
of plain rollers above
rocking rollers $= 24 \text{ in.} + 13 \text{ in.}$
$- 22\frac{1}{2} \text{ in.}$ $= 14\frac{1}{2} \text{ in.}$

(2) Downward tilt of
launching nose from
difference in ground
level $= \dfrac{125 \text{ ft.}}{75 \text{ ft.}} \times 14\frac{1}{2} \text{ in.} = 24 \text{ in.}$

Difference between
ground level of far- and
near-bank seats $= 36 \text{ in.}$

Sag for 130-ft. double-
double bridge $= 30 \text{ in.}$

Therefore, the resulting
sag $= 90 \text{ in.}$

(3) From table XIX, it is found that where the sag of the tip
of the launching nose is 90 inches, links must be used at 30 or
40 feet from tip of launching nose (see fig. 66).

Note: In launching the bridge, the lowest point of tail, (*A*),
is when (*B*), rear launching nose link, reaches far bank. (*A*)
then is 35 feet behind rocking rollers on near bank. Neglect-
ing effect of sag, which will have little influence, (*A*) then will
approximately be $\dfrac{35 \text{ ft.}}{125 \text{ ft.}} \times 36 \text{ in.}$, or 10 in. below top of
rocking rollers on near bank. Therefore, for (*A*) to clear near
bank, ground level 35 feet behind rocking rollers must not
exceed $22\frac{1}{2}$ in. — 10 in., or $12\frac{1}{2}$ in. above the ground level
at the bank seat.

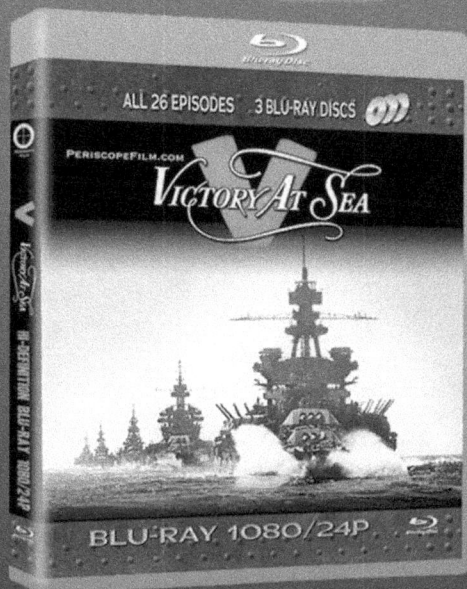

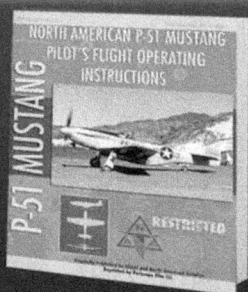

www.ingramcontent.com/pod-product-compliance
Lightning Source LLC
Chambersburg PA
CBHW071116210326
41519CB00020B/6323